BRAIN–COMPUTER INTERFACING FOR ASSISTIVE ROBOTICS

ELSEVIER *science & technology books*

Companion Web Site:

http://booksite.elsevier.com/9780128015438

Brain–Computer Interfacing for Assistive Robotics
Vaibhav Gandhi B.Eng, M.Eng, Ph.D

Resources available

• Videos highlighting the intelligent Adaptive User Interface (iAUI)

TOOLS FOR ALL YOUR TEACHING NEEDS
textbooks.elsevier.com

**ACADEMIC
PRESS**

BRAIN–COMPUTER INTERFACING FOR ASSISTIVE ROBOTICS

Electroencephalograms, Recurrent Quantum Neural Networks and User-Centric Graphical Interfaces

VAIBHAV GANDHI B.ENG, M.ENG, PH.D
School of Science and Technology, Middlesex University, London, UK

AMSTERDAM • BOSTON • HEIDELBERG • LONDON
NEW YORK • OXFORD • PARIS • SAN DIEGO
SAN FRANCISCO • SINGAPORE • SYDNEY • TOKYO

Academic Press is an imprint of Elsevier

Academic Press is an imprint of Elsevier
32 Jamestown Road, London NW1 7BY, UK
525 B Street, Suite 1800, San Diego, CA 92101-4495, USA
225 Wyman Street, Waltham, MA 02451, USA
The Boulevard, Langford Lane, Kidlington, Oxford OX5 1GB, UK

ISBN: 978-0-12-801543-8

British Library Cataloguing-in-Publication Data
A catalogue record for this book is available from the British Library

Library of Congress Cataloging-in-Publication Data
A catalog record for this book is available from the Library of Congress

For information on all Academic Press publications
visit our website at http://store.elsevier.com/

Typeset by MPS Limited, Chennai, India
www.adi-mps.com

Printed and bound in the United States of America

Working together
to grow libraries in
developing countries

www.elsevier.com • www.bookaid.org

Contents

List of Figures

List of Tables

Preface

Brain—computer interface (BCI) technology provides a means of communication that allows individuals with severely impaired movement to communicate with assistive devices using the electroencephalogram (EEG) or other brain signals. The practicality of a BCI has been made possible by advances in multi-disciplinary areas of research related to cognitive neuroscience, brain-imaging techniques and human—computer interfaces. The end goal of a BCI is to enable monitoring of the underlying brain processes and subsequent utilization of this information for communicating and controlling devices solely through the brain without depending on the normal output pathways of peripheral nerves and muscles.

This book is a result of four years of doctoral research, focusing on an important aspect of BCI for real-time assistive robotic application. This book will benefit both mature and early-stage researchers, including postgraduate and doctoral students. Researchers from other fields interested in BCI should also find this book interesting and informative, including its detailed discussion of prevailing research trends. This book should also be helpful in courses related to quantum mechanics, neural networks, signal processing, robotics and human—computer interaction, since it covers the fundamental postulates of quantum mechanics, their application for filtering noisy EEG signals, signal processing fundamentals and adaptive graphical user interfaces (GUIs) for robot control application in considerable detail.

Dealing with the unknown embedded noise within the raw EEG and the inherent lower bandwidth of BCI are still two of the major challenges in making BCI practical for day-to-day use. The raw EEG signal recorded non-invasively during motor imagery (MI) is intrinsically embedded with non-Gaussian noise, and the actual, noise-free EEG has so far not been obtained. Therefore, a novel filtering approach utilizing the concepts from quantum mechanics is discussed in this book. This filtering approach is constructed using a layer of neurons within the neural network framework and is referred to as the Recurrent Quantum Neural Network (RQNN).

Another major challenge in two-class BCI systems is the inherently low bandwidth of the communication channel, which may lead to a sluggish response when controlling a mobile robotic device. An intelligent and adaptive user interface, which plays a very important role as a front-end display for the BCI user, is discussed in this book. The framework of the proposed intelligent Adaptive User Interface (iAUI), also referred to as a brain−robot interface, is consistent for use with a range of applications, e.g., for controlling either a mobile robot or a robotic arm. Both the RQNN filtering technique and the iAUI have been investigated for real-time application and are discussed in detail in this book.

The book contains seven chapters. Chapter 1 introduces the rationale, the aims and objectives of the book. Chapter 2 presents in detail a general EEG-based BCI system. The chapter begins with an introduction of the biological fundamentals of the brain and a description of EEG rhythms. This is followed by a description of the operational techniques in BCI (synchronous and asynchronous communication) and an in-depth explanation of the various signal processing related building blocks of a typical BCI system including the data acquisition process, preprocessing, feature extraction, classification and post-processing approaches. Cross-validation and a few popular optimization techniques such as particle swarm optimization (PSO) and genetic algorithms (GA) which are utilized for tuning/selecting the RQNN model parameters are discussed. The chapter then describes the necessity and importance of a GUI, the different interface designs proposed by various research groups and the challenges and issues for practical implementation. Major BCI strategies and breakthrough techniques employed to enhance the usability of BCI signal processing outcome for applications in the form of shared control and adaptive shared control techniques are also discussed along with the performance evaluation measures for an interface. Chapter 3 begins with a brief overview of the fundamentals of QM and SWE. This chapter then describes the theoretical concept of the RQNN models and their application to filtering simple signals in the form of DC, stepwise DC and sinusoidal waveforms.

The proposed MI-based GUI design, i.e., the adaptive interfaces for a wheelchair/robot control application, are discussed in Chapter 4. This chapter begins by discussing the role of the proposed brain−robot in the complete BCI framework, and then covers the mobile control and arm control applications. Chapter 5 builds on the foundations laid in the form of the proposed quantum mechanics-based RQNN models discussed in Chapter 3. The traditional and the revised RQNN models are investigated for different BCI competition datasets, as are subject-specific RQNN model parameters using a two-step inner-outer five-fold cross-validation technique. Chapter 6 presents the brain−robot interface for a practical robot control task. The EEG acquisition process within the BCI laboratory and the procedure for applying the RQNN technique in real time are also discussed. The proposed autonomous and supervised interfaces for real-time maneuvering of the mobile robot to reach different locations within the simulated and physical robotic arena are then discussed and compared with existing designs in the field of BCI. Chapter 7 concludes the book, including potental future directions and research ideas to potentially further enhance the performance of the BCI system. References and appendices are given at the end of the book.

Vaibhav Gandhi

Acknowledgments

This book is based on Vaibhav Gandhi's Ph.D. thesis, which was submitted to the University of Ulster, UK. The research work was carried out at the Intelligent Systems Research Center, University of Ulster, UK and the Department of Electrical Engineering, IIT Kanpur, India. The author is thankful to his family and friends whose personal support has been instrumental in making this book possible.

List of Acronyms

1-D	1-dimensional
ADC	Analog to Digital Converter
ALS	Amyotrophic Lateral Sclerosis
AM	Adaptation Module
ANN	Artificial Neural Network
ANOVA2	Two-way Analysis of Variance
AR	Autoregressive
ARM	Autonomous Arm
BCI	Brain–Computer Interface
BCW	Brain-Controlled Wheelchair
BP	Band Power
CA	Classification Accuracy
CAR	Common Average Reference
CM	Communication Module
CMRR	Common-Mode Rejection Ratio
CSCI	Cervical Spinal Cord Injuries
CSP	Common Spatial Patterns
CSSP	Common Spatio-Spectral Pattern
CSSSP	Common Sparse Spectral Spatial Pattern
DFT	Discrete Fourier Transform
ECG	Electrocardiogram
EEG	Electroencephalogram
EKF	Extended Kalman Filter
EMD	Emperical Mode Decomposition
EMG	Electromyogram
EOG	Electrooculogram
EP	Evoked Potential
ERD	Event-Related Desynchronization
ERP	Event-Related Potential
ErrP	Error Potential
ERS	Event-Related Synchronization
FBCSP	Filter-Bank CSP

FE	Feature Extraction
FEP	Feature Extraction Process
FLDA	Fisher's LDA
FT	Fourier Transform
GA	Genetic Algorithm
gbest	global best
GUI	Graphical User Interface
hBCI	hybrid BCI
HHT	Hilbert–Huang Transform
iAUI	Intelligent Adaptive User Interface
ICA	Independent Component Analysis
ID	Index of Difficulty
IMF	Intrinsic Mode Functions
IP	Index of Performance
IRLS	Iteratively Reweighted Least-Squares
IRM	Information Refresh Module
ITR	Information Transfer Rate
KF	Kalman Filter
lbest	local best
LDA	Linear Discriminant Analysis
MI	Motor Imagery
MLE	Maximum Likelihood Estimation
MM	Monitor Module
MND	Motor Neurone Disease
MT	Movement Time
NC	No-Control
NN	Neural Network
NTSPP	Neural–Time-Series-Prediction
OVR	One Versus Rest
PCA	Principal Component Analysis
pdfs	Probability Density Functions
PMs	Prediction Models
PSD	Power Spectral Density
PSO	Particle Swarm Optimization
QM	Quantum Mechanics
QNN	Quantum Neural Network
RBF	Radial Basis Function

RCSP	Regularized CSP
RMSE	Root Mean Square Error
RQNN	Recurrent Quantum Neural Network
SARM	Supervised Arm
SBCSP	Sub Band CSP
SCI	Spinal Cord Injury
SCP	Slow Cortical Potentials
SG	Savitzky–Golay
SMA	Supplementary Motor Areas
SMR	Sensorimotor Rhythm
SNR	Signal-to-Noise
SSVEP	Steady State Visually Evoked Potential
SVM	Support Vector Machines
SWLDA	Stepwise LDA
t-f	Time-Frequency
TSD	Time-Varying Signed Distance
UDP	User Datagram Protocol
UKIERI	UK-India Education & Research Initiative
UMDA	Univariate Marginal Distribution Algorithm
VB	Visual Basic
VEPs	Visually Evoked Potentials

Introduction

1.1 INTRODUCTION

Verbal or non-verbal information exchange is the basis of human communication. However, some people lose this fundamental ability of communication because of accidents or inherited neuromuscular disorders. The purpose of the work presented in this book is to contribute to the development of novel methods to allow people to regain freedom of movement/communication by way of controlling devices directly with their brain, bypassing the normal communication channels.

The human brain is estimated to contain about 100 billion neurons [1–4]. The spinal cord acts as an intermediate cable that carries information to and from our brain to control various body parts and their movements. People with an injury to the spinal cord are still able to generate the output signals from the brain, but these signals do not reach the specific body parts because the intermediate spinal cable is damaged. Several technologies using a joystick, head movement, eye gazing and many more may help a physically challenged person to control a robotic device or a wheelchair [5–9]. However, these techniques require the use of partial movement control through the hand, head or eyes etc., and therefore make the control issue less complicated. The issue becomes more challenging when people with complete loss of control over their voluntary muscles are involved, a condition generally known as locked-in syndrome [10,11], in which people are unable to speak and move but are conscious and can think and reason. A number of neurological diseases such as stroke[1], severe cerebral

---

[1]Stroke is a brain injury caused by an abnormality of the blood supply to a part of the brain.

1

palsy[2], motor neuron disease (MND)[3], amyotrophic lateral sclerosis (ALS), and encephalitis[4] can result in such severe motor paralysis [12]. Many of these diseases can lead to restrictions in communication capacity. A brain–computer interface (BCI) can enable such physically challenged people to achieve greater independence by making technology accessible. BCI technology provides an alternative communication channel between the human brain (that does not depend on the brain's normal output channels of the peripheral nerves and muscles) and a computer [13–21]. The three most commonly discussed diseases/injuries cited in the BCI literature as being a case of locked-in syndrome are ALS, high spinal cord injury and brain stem stroke [16,22–24].

- Patients suffering from ALS can undergo severe physical impairment due to the degeneration of nerve cells that control the voluntary muscles. In the later stages of ALS, the most basic human actions are affected, including speech, swallowing and breathing [25].
- Spinal cord injury (SCI) can result in damage to myelinated fiber tracts or the nerve roots that carry the signals to and from the brain [25]. In complete SCI, most of the motor functions and sensation below the neurological level are affected or completely lost [26]. SCI has a global annual incidence of 15–40 cases per million population [27] and less than 5% of people suffering from SCI recover locomotion [26].
- Brain stem stroke can be fatal, as the brain stem controls many of the basic and fundamental activities for life, such as breathing, heart rate, blood pressure, swallowing and eye movement [28]. People with severe brain stem stroke may also enter into a locked-in state and lose motor functions [29].

BCI (i.e., electroencephalography [EEG])-based communication produces new channels for controlling devices which would not be possible through the modes of communication that require eye movement or some muscle activity. Hans Berger performed a systematic study of the

---

[2]Severe cerebral palsy is a non-progressive, but not unchanging, disorder of movement and posture that is the consequence of lesions or anomalies of the brain arising in the early stages of its development [365].

[3]MND actually describes a group of very similar conditions that affect motor neurons. Amyotrophic lateral sclerosis (ALS) is the most common type of MND [366] and affects both the upper motor neurons (those running from the brain to spinal cord) and the lower motor neurons (those running from the spinal cord out to the muscles). ALS is also referred as Lou Gehrig's disease in the USA and MND in the UK [367].

[4]Encephalitis is an inflammation of the brain, usually from an infection, but most often by a virus and sometimes by bacteria, fungi or parasites.

electrical activity of the human brain, and developed the EEG[5]. The first scientific literature referring to communication between the brain and the computer dates back to the early 1970s, and is due to Vidal [18], who suggested the feasibility of direct brain communication. To achieve this, the intent of the user must be extracted from the brain via the EEG or brain waves.

A typical BCI scheme generally consists of a data acquisition system, preprocessing of the acquired signals, the feature extraction process (FEP), classification of the features and finally the control interface and device controller, as shown in Figure 1.1. The EEG signals are acquired by mounting electrodes on the scalp of the user. These raw EEG signals have very low amplitude [30], very low signal-to-noise (SNR) ratio and considerable noise contamination. Preprocessing is carried out to obtain cleaner EEG signals by removing the unwanted components embedded in the EEG, which can considerably reduce the computational load on the rest of the BCI components.

The work presented in this book has focused on the preprocessing stage for signal enhancement and extracting more motor imagery (MI)[6] (mental imagination of movement) [31] related information from the acquired noisy EEG. These raw EEG signals are considered as a realization of a random or stochastic process [32]. When an accurate description of the signal is not available, a stochastic filter can be designed

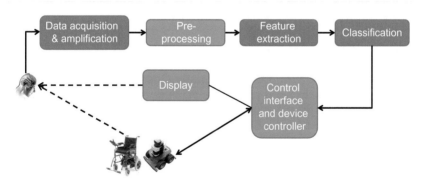

FIGURE 1.1   Basic functional block diagram of a simple BCI system.

[5]Berger made the first recording of the EEG in man in 1924 and named it the 'Electroenkephalogram'. The historical progression of EEG, EEG instruments and techniques is presented in [368] by Collura.

[6]MI is the mental process by which an individual simulates a specific motor action within the so-called internal imagination state without any actual physical implication.

based on probabilistic measures. Therefore, the approach undertaken in this book is to use the concepts from quantum mechanics (QM) and the Schrodinger wave equation (SWE). A recurrent quantum neural network (RQNN) is constructed by using a layer of neurons within the neural network framework by computing a time-varying probability density function (*pdf*) of the noisy input signal (cf. Chapter 3). This *pdf* evolves recurrently under the influence of the SWE and helps to enhance the EEG (cf. Chapter 5).

Features are extracted from the filtered EEG. The classifier interprets these extracted features to categorize the input signal into a designated output class. The control interface or the graphical user interface (GUI) further interprets the classifier output in the form of a command to be sent to the controlled device. The GUI also provides appropriate feedback information to the BCI user [7], and can quicken the issuance of the command from the BCI user to the device that is controlled. A two-class BCI system has two output classes in the form of a left hand MI or a right hand/foot MI. But the task of maneuvering a mobile robot requires commands in the form of forward, left, right, backward and start/stop, using just the two-class information; i.e., there is very limited communication bandwidth. However, given the inherent higher accuracy compared to multi-class BCIs, this book focuses on utilizing a two-class BCI. This book proposes an intelligent, adaptive and user-centric interface design that plays a major role in compensating for the low bandwidth of a two-class BCI and simultaneously capitalizes on the intrinsic higher accuracy characteristic that is typical of a two-class BCI system (cf. Chapter 4 and Chapter 6).

In summary, this book outlines the recent developments in MI-based BCI, specifically focusing on reviewing the existing signal processing and classification methodologies, as well as different interface designs for a BCI system. It proposes an alternative nature-inspired information processing approach based on the concepts from QM, which is referred to as the RQNN model and is utilized for EEG signal enhancement (cf. Chapter 3 and Chapter 5). It also proposes an intelligent user interface design (cf. Chapter 4) which is customized to provide effective control of a wheelchair/mobile robot and a robot arm for a quicker communication process (cf. Chapter 6).

1.2 RATIONALE

BCI technology has not yet reached a critical level of acceptance even forty years after its inception. The challenges in this domain begin right at the stage of acquiring the EEG signals from the brain.

An EEG is recorded non-invasively, so it is a mixture of the signal of interest from the activity of the underlying neural networks and an unknown amount of noise. Therefore, the raw signals need to be filtered in order to obtain cleaner EEG signals. Several groups work in the field of EEG filtering [19,33−39]. Most of their approaches involve subject-specific parameters, which, if tuned properly, can enhance the performance of an individual subject in terms of the classification accuracy (CA) [35,40]. However, these frequency-selective techniques lead to an unknown amount of loss of information from the acquired EEG signals [41,42]. Therefore, the frequency-selective EEG filtering methods pose challenges for long BCI setup and adaptability [43].

A major challenge in the area of EEG filtering is that the actual noise-free EEG is still unattainable. The work presented in this book addresses to some extent the very intriguing and challenging question of working towards a noise-free EEG, by relying heavily on the principles of QM to produce a time-varying *pdf* of the stochastic EEG signal, which evolves under the influence of the SWE to produce a cleaner EEG. It needs to be emphasized here that it can only be assumed that a cleaner EEG has been obtained if certain performance parameters (such as the CA of the BCI) are better than the unfiltered/raw EEG after performing the filtering process. Appropriate features or attributes [19,38,39,44−54] are obtained from these enhanced EEG signals in order to make the discrimination task easier at the classifier stage. However, a major challenge in two-class BCI systems is the inherent lower bandwidth of the communication channel, which may lead to a slower response when controlling a robotic device. This study has attempted to address this challenge by creating an adaptive and dynamic interface design that is updated as the environment surrounding the controlled device changes. This approach resolves the issue of lower bandwidth of the BCI to some extent by improving the communication speed of the system, and enabling the BCI user to reach the most anticipated choice quickly, because it is displayed as the most accessible option [55−57] (cf. Chapter 4 and Chapter 6).

As far as the existing developments in BCI are concerned, a practical brain-controlled wheelchair/robotic system is as yet not feasible because of some of the above issues, which still prevent guaranteed mental imagination detection. In addition, a practical BCI process would require the system to be accurate in real time. The next section details the objectives that have been set to resolve some of the above issues, focusing on noise removal and a dynamic and adaptive interface design to enhance the overall communication process.

1.3 OBJECTIVES

To address the major challenges of dealing with the unknown embedded noise within the raw EEG and the inherent lower bandwidth of BCI, to make BCI practical for day-to-day use, the main objectives of this book are:

- To investigate QM concepts and obtain ideas/motivation for applying QM-based fundamentals as an EEG filtering methodology for a BCI. QM, in general can be considered as a mathematical approach that describes the behavior of particles at the microscopic level, but the QM applied in the quantum neural network framework is conceptual and the approach does not work at a particle or microscopic level.
- To investigate the RQNN-based preprocessing methodology for filtering the raw EEG signal. The incorporation of SWE into the neural network paradigm provides a structure that is referred to as RQNN. The SWE is the fundamental equation in QM for describing quantum mechanical behavior, and it is analogous to Newton's second law of motion in classical mechanics.
- To investigate the performance of the RQNN filtered EEG by utilizing different features from the time and frequency domains and performing the classification process by using different classifiers.
- To investigate a post-processing methodology for appropriately tackling the issue of ambiguous/disputed classifier outputs within a single EEG trial. The approach undertaken in the present work is to utilize the concepts of de-biasing and multiple threshold methodologies.
- To investigate a practical, synchronous and shared control based adaptive brain−robot interface paradigm (i.e., the GUI) designed to run in tandem with the proposed RQNN filtering technique and the existing FE and classification techniques. As mentioned above, a user-centric brain−robot interface that can adapt itself in real time to the available options and most probable choices from the BCI user is sought, because it can make best use of the limited bandwidth of the communication process. This book aims to address the limited bandwidth issue of a two-class BCI without adding any complexity by using novel post-processing methodology and an innovative brain−robot interface.

2

Interfacing Brain and Machine

2.1 INTRODUCTION

In simple terms, a brain—computer interface (BCI) is a direct interface between the human brain and an artificial system. Its purpose is to control the actuation of a device, say a robotic system or a wheelchair, with brain activity but without the use of peripheral nerves or muscles [19]. BCI in a literal sense means interfacing an individual's electrophysiological signals with a computer [58]. Thus, in a true sense, the BCI only uses signals from the brain and must consider eye and muscle movements as artifacts[1] or noise. Information from various knowledge domains is necessary to create a complete BCI system.

This chapter reviews the work implemented by various interdisciplinary groups in the development of a BCI system along with the different modes of operation, and begins with an introduction to the fundamental biological aspects of BCIs.

2.2 THE BRAIN AND ELECTRODE PLACEMENT

The adult human brain houses approximately 100 billion neurons [1—4] which are responsible for generating the electrical activity that drives a typical electroencephalogram (EEG) signal for the BCI system. It is said that sometimes the forest is more important than the trees. In this case, one is often less concerned with the activities of individual neurons than with understanding the activity of a large population of neurons [59]. It has been shown that groups of neurons with similar

---

[1]Artifact is an unwanted signal detected along with the EEG but originates from non-cerebral origin.

7

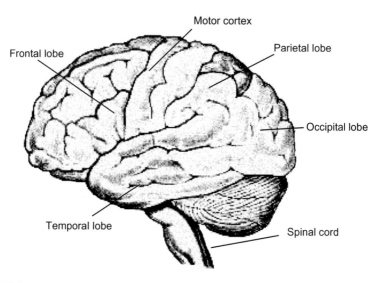

FIGURE 2.1   Coarse overview of some important brain areas.

functions fire in the same location on the cerebral cortex. This gives rise to a functional map of the brain as shown in Figure 2.1. The main gross brain areas [60] are the frontal, parietal, occipital and temporal lobe regions [59]. The motor cortex is located in the rear portion of the frontal lobe, just before the central sulcus that separates the frontal lobe from the parietal lobe (a narrow strip near the top of the head). This area of the brain is responsible for controlling movement. Thus, if electrodes are placed over the motor cortex, EEG signals associated with motor imagery (MI) can be acquired. MI can be interpreted as the mental rehearsal of a motor act [61,62]. If one wants to measure the P300 signals (a type of event-related potential; cf. Section 2.4.2 for more detail) that dominate over the central vertex region, or the visually evoked potentials (VEPs; cf. Section 2.4.2 for more detail) which dominate the occipital region, then electrodes should be mounted accordingly. Understanding of the structure of the brain is therefore important for informing the placement of electrodes on the scalp of the BCI user. If the optimal recording sites are correctly associated with the command-related EEG activity then the probability of success of the BCI may be increased. Most of the present day BCI systems are non-invasive, i.e., the mounting of the electrodes is external to the scalp. This is accomplished by wearing an electrode cap such as the one shown in Figure 2.2 [63], which helps to obtain the EEG signals from the scalp. This approach is preferred, as it does not require surgical implantation of electrodes inside the brain by invasive means and makes the BCI practically usable and user friendly.

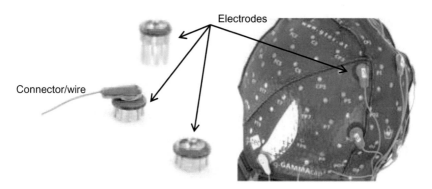

FIGURE 2.2   Mounting of electrodes using the dry EEG cap. *Reproduced with permission from g.tec medical engineering GmbH. Copyright 2012.*

2.2.1 EEG Wave Rhythms

The human brain continuously generates minute electrical activities in the form of voltage signals and these show fluctuations roughly in the frequency[2] range from 0.5 Hz to 100 Hz, but fundamentally in the range from 0.5 Hz to about 50 Hz [15]. This range can further be divided into different bands of frequencies [15,64,65] and correlated to different cognitive states or movements, as indicated in Table 2.1. This division arises because it has been found that each of these bands has a characteristic biological significance.

The delta band is normally in the frequency range of 0.5 to 4 Hz and is associated with deep sleep [66]. These rhythms are normally very slow waves but are the highest in amplitude. It is worth noting here that with these slower dominant frequencies, the responsiveness to stimuli also decreases [67]. The theta band is in the range of 4 to 7 Hz and is dominant in children during rest or sleep. It is also observed in adults and older children during drowsiness and also in the state of performing meditation or being in a relaxed or creative state. The alpha band is in the range of 8 to 13 Hz, and is commonly observed in the posterior and occipital regions of the human brain. Healthy subjects often reveal a predominance of the alpha rhythm when the eyes are closed [15]. Opening the eyes causes a blocking of the alpha band, while the other frequencies usually become more obvious. These alpha waves are the most commonly studied rhythms of the human brain. Mu (μ) rhythms are a special kind of normal alpha rhythm that are observed within the range of 9 to 11 Hz. They are almost always present when the subject is

---

[2]Prawdicz-Neminski (1925) was the first to classify the brain's electrical activity in terms of frequency.

TABLE 2.1 Brain Rhythms and Relevant Frequency Band Details

Brain rhythm	Typical frequency range (Hz)	Normal amplitude (μV)	Comments
Delta	0.5–4	<100	Dominant in infants and during deep stages of adult sleep. Found in the central cerebrum and parietal lobes
Theta	4–7	< 100	In children and drowsy normal adults Found in the frontal, parietal and temporal regions
Alpha	8–13	20–60	This is the most prominent rhythm in the normal alert adult brain. Most prominent in the occipital and parietal regions
Mu	9–11	<50	This frequency band is associated with hand movements. Found over the motor and somatosensory cortex
Beta	14–30	<20	This frequency band is also associated with hand movements
Gamma	>30	<2	Found when the subject is paying attention or is having some sensory stimulation

relaxed, and disappear when the subject moves a hand or finger on the contra-lateral side; i.e., the μ rhythms disappear over the left brain hemisphere when the right hand is moved and vice versa [15]. Thus this frequency band can be associated with hand movement (and also with the MI of the hand movement) and can be used to distinguish between movements of either of the hands. The beta rhythm band is observed in the range of 14 to 30 Hz, and normally has a symmetrical distribution on both sides of the brain although they are more dominant on the frontal side. These rhythms show synchronization in the ipsi-lateral side during active movements, hence this band can also be utilized for hand movement detection. The gamma rhythm bands are in the wide range of 30 to 100 Hz and display cross modal sensory processing [68]; i.e., interaction between two or more different sensory modalities. Recent investigation provides evidence to suggest a relationship of EEG recordings in the gamma range of 55–85 Hz with the sensorimotor rhythm (SMR) [69]. It has also been demonstrated that modulation of the SMR, induced by MI of either the left or right hand, is correlated with gamma oscillations [70]; in other words, there is an underlying influence of gamma oscillations on the SMR. As modulation of the SMR is typically used in BCIs to infer a subject's intention, these findings mean that gamma oscillations also have a causal influence on a subject's capability to utilize a BCI for communication. All the frequency bands mentioned above are approximate and should be used as a guideline only. The ranges of these different bands also vary from one subject to another.

Another important point to discuss here is that the contribution of each of the frequency bands to the overall EEG curve depends on the specific situation. Dastidar [71] states that these individual frequency sub-bands may be more representative of the brain dynamics than the entire EEG and also makes the point that the sub-bands gave more information about the underlying neuronal activities.

2.3 OPERATIONAL TECHNIQUES IN BCI

BCI systems are categorized as two modes;

- Synchronous BCI, which is a cue-based and computer-driven approach; and
- Asynchronous BCI, which is a non-cue-based, user-driven approach, also called self-paced BCI.

2.3.1 Synchronous BCI

In synchronous BCI systems, the subject acts after receiving a visual or auditory cue stimulus generated by the computer. A trial may normally include a sequence of two events: first is the display of a cue informing the subject to be ready, and second, after a predefined fixed time interval of a few seconds, a further cue informing the subject to perform the desired mental task. Thus, it is a cue-based and computer-driven system. The timing and the control commands are generated by the computer system and the user should act as per the timing generated by the computer. The advantage of synchronous BCI systems is that the onset of mental activity is known in advance and associated with a specific cue or trigger stimulus, and thus any signal outside the predefined time windows is treated as idling and ignored by the BCI system [72]. This makes the synchronous BCI the best choice for most of the BCI groups. Here, the system is initially trained offline using a synchronous protocol with a specific time interval for a particular subject. A typical trial of a synchronous BCI system lasts for about 4–10 s including the actual mental imagination/MI time duration, the duration of rest and the inter trial time period. The EEG phenomena to be recognized are time locked to the duration when the user performs the desired mental task for a certain predefined length of time. The activity of millions of neurons must be synchronized and desynchronized to develop the desynchronization and synchronization of alpha (μ) rhythms (cf. Section 2.4.2) respectively, which need a period time of the order of seconds to develop [73]. Therefore, a long time is needed for the EEG phenomena of interest to recover from the previous brain state [74], associated with the preceding imagination trial. Another advantage of using the synchronous approach is that the BCI user performing the MI gets time to take rest in between the trials because the user needs to perform the MI only for a duration of approximately 4 s during the complete cycle of a single trial. Conversely, it can also be said that this system also has the inherent drawback that the command can be issued only once in a trial. Hence, the overall communication bandwidth of these systems is limited by the length of the predefined trial.

Figure 2.3 shows a typical synchronous BCI paradigm. The training session involves sequential but random repetition of the cue-based trials. The trial begins with a fixation cross that is displayed in the center of a monitor. After two seconds, a warning stimulus is presented in the form of a 'beep.' This is to inform the user to get ready to perform the MI. From second 3 to second 4.25 an arrow (cue stimulus) pointing to the left or to the right is shown on the monitor. The subject then has imagine movements as indicated by the cue on the screen (depending on the direction of the arrow) until second 8. As mentioned before, this duration of 8 s is typical of a single trial. The order in which these trials are presented to the user is generated randomly. In addition, the time between any two trials is also randomized in a range of 0.5 to 2.5 s. Randomness of presenting time of the heterogeneous cue raises the power of concentration of the subjects, and this is thought to be the main cause of the consistency in performance [75]. The subject is normally required to perform four runs, each consisting of 40 trials. Thus, a single session of a two-class MI typically consists of 160 trials. This type of arrow paradigm has been followed by several research groups for offline analysis [48,76–78].

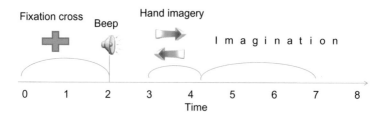

FIGURE 2.3   Training paradigm and timing.

2.3.2 Asynchronous BCI

Asynchronous BCI systems are non-cue-based and user-driven. This should provide a more natural way of communication between the user and the device to be controlled. However, the approach requires continuous analysis and feature extraction of the recorded brain signals. The major drawback of these systems is that the user has to be completely attentive in performing the MI at all times during the complete session. If the user becomes non-attentive even for a few seconds it can lead to a different and a wrong command being sent to the controlled device and this may result in unexpected consequences. Another challenge with asynchronous systems is the difficulty in clearly distinguishing the no-control (NC) state (in which the user intends not to alter any of the control outputs) when the user does not wish to issue any command. This is also referred as the inactive period (idle state). Thus, in these systems, it is a major challenge to accurately associate this intention/no intention with a specific control objective because the user performs the mental tasks at his/her own will and timing. The continuous classification of the user's brain activity for achieving asynchronous interaction can cause random false positives (i.e., incorrectly identifying an idle state as an intentional control) during periods when the user is not executing any MI task. Thus, it is also a major challenge to accurately distinguish the true positives (i.e., correctly identifying an intentional control) and the false positives. Only a limited amount of work has been done with these systems, although there have been a number of recent developments [22,79−82].

Results from various studies support the notion that performing an MI and executing the task involves activation of very similar cerebral structures 'at all stages of motor control' [83] except the final motor output[3]; i.e., the actual movement of the body part is not expressed while performing the MI [84,85]. Thus, a person not affected by any brain or spinal injury can produce the motor signals by imagining a movement without actually moving the body parts [85]. For a healthy subject, imagining the movement without performing it is quite strange and unnatural. Nevertheless, this also means that before experimenting with physically challenged subjects, a BCI system can be tested on able-bodied subjects. In the work presented in this book, the focus is on the mental states for synchronous MI-BCI design; i.e., identifying the MI of the left hand/right hand/foot.

---

[3]This notion of equivalence is well-evidenced [369−372], however, some characteristics of MI remain largely unexplored [373].

2.4 DATA ACQUISITION

2.4.1 The Basics of Data Acquisition

The neurons of the cerebral cortex generate varying electrical fields on the surface of the skull that can be recorded with appropriate electrodes. The summated activity of the neurons can be registered as an EEG and projected onto the area of the recording electrode. The potential changes on the cortical surface largely depend on the postsynaptic potentials[4] at the dendrites[5] of the pyramidal neurons[6] (cf. Figure 2.4). Since the pyramidal neurons are positioned at right angles to the cortical surface, they have much greater impact on the surface potential than other neurons. Further, these pyramidal neurons are all orientated in parallel to one another, so the equidirectional potential changes of the neighboring pyramidal neurons are summated. When several pyramidal neurons are simultaneously excited, an EEG deflection can be expected [86]. The EEG thus represents a very small potential of approximately $10\,\mu V$ to $100\,\mu V$ when measured on the scalp and $1-2\,mV$ when measured on the surface of the brain [87]. To access this potential, most BCI researchers mount electrodes on the scalp of an individual (cf. Figure 2.2). The positioning of the electrodes is in accordance with the 10/20 electrode placement system [88] as shown in Figure 2.5 (reproduced from [89]). This is based on the relationship between the location of the electrode and the underlying area of the cerebral cortex, as discussed in Section 2.2. Thus, this system defines the location of the electrodes externally on the scalp. Three anatomical reference points should be determined before the electrodes are mounted on the skull of the user. These reference points are the positions of the nasion, inion and the preauricular reference point as shown in Figure 2.5(a). The nasion is a region below the forehead and at the onset of the nose on the skull. The inion is a small bony protrusion between the skull and the neck. And the preauricular reference point is located before the cartilaginous protrusion of the auditory canal. Each of the electrode locations in the 10/20 system is marked by a letter and a number. The letter identifies the lobe, and the number identifies the hemisphere that the electrode belongs to. F stands for frontal, T for temporal, C for central, P for parietal, O for occipital and A for auricular. A 'z' is used to refer to the electrode placed on the midline. Even numbers mean that the electrode is positioned in the right

---

[4]Postsynaptic potential is a temporary change in the electric polarization of the membrane of a neuron.

[5]Dendrite is a short branched projection of the neuron.

[6]Pyramidal neurons (have pyramidal shaped body) are found in the forebrain structure and play an important role in advanced cognitive functions.

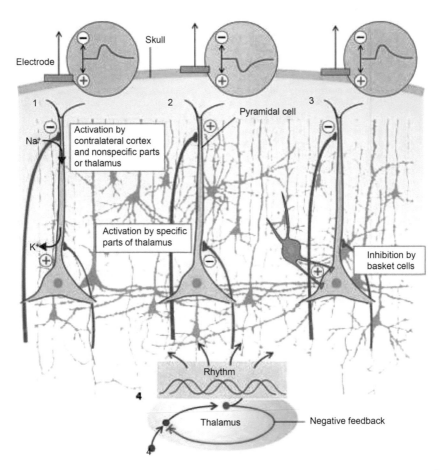

FIGURE 2.4   Genesis of the EEG. *Reproduced with permission from [86].*

part of the hemisphere, while odd numbers refer to positions in the left part of the hemisphere. Once the placement of the electrodes is understood, the next task is to record the EEG.

For recording purposes, electrodes can be connected in either unipolar or bipolar mode, as shown in Figure 2.6. In unipolar mode, one of the electrodes is connected to the positive input of the amplifier. The other electrode, which offers the return current to the negative input of the amplifier, is attached to a common or reference position. Thus in unipolar mode, all the negative electrodes are connected to the same common or reference electrode. The potential of each electrode is compared either to a neutral electrode or to an average of all the electrodes. In the bipolar configuration, both the electrodes are connected to

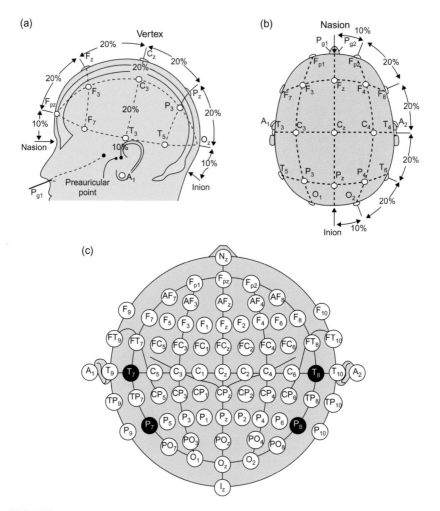

FIGURE 2.5 The extended 10–20 system for electrode placement. *Reproduced with permission from [89].*

specific locations over the scalp as shown in Figure 2.7. Thus in this case the potential difference between a pair of electrodes is measured. More details about the guidelines for EEG recording can be found in reference [90].

A single action potential or neuronal response cannot be registered at the surface of the scalp, and hence any potential change that is measured by the EEG recording is the effect of thousands of neurons firing simultaneously. The potential difference signals (voltages) from an adult human brain's electrical activity are very small; typically the usual amplitude of EEG oscillations is in the range of 10 to 100 μV [91].

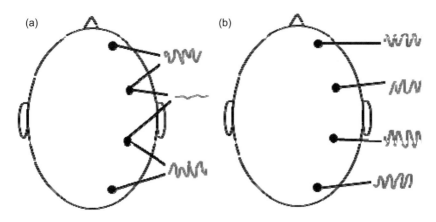

FIGURE 2.6   (a) Bipolar measurement. (b) Unipolar measurement.

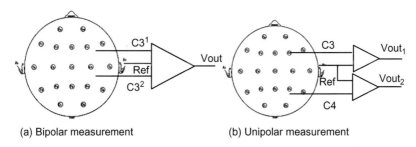

(a) Bipolar measurement   (b) Unipolar measurement

FIGURE 2.7   EEG measurement.

Hence the impedance between the scalp and the electrodes is a key factor in ensuring that these low-amplitude signals pick up the least amount of external noise. Normally, the impedance between the electrodes and the scalp should be kept below 10 kOhms, in order to reduce the noise levels of the EEG [92,93]. Impedances of around 5 kOhms are normally considered optimal for EEG signal recordings. The EEG signals picked up by the EEG electrodes are then amplified (usually 1,000–100,000 times, or 60–100 dB of voltage gain) [94] in order to make them displayable and discriminable. The important function of an amplifier is to pick up the weak biological signals and increase their amplitude so that they can be further processed. Amplifiers in clinical neurophysiology provide an amplified version of the difference of the two potentials at the input to the amplifier [95]. These differential amplifiers cancel the unwanted signals which are identical at both the inputs. This is known as the common-mode voltage and is responsible for a lot of interference in the amplifiers. An amplifier with a higher common-mode rejection ratio (CMRR)[7] minimizes the effects of these common-mode voltages, however a better way to approach noise issues is to determine the source of the voltage and try to eliminate it, e.g., using electrostatic shielding or grounding to reduce electric and magnetic interference. The EEG signals picked up from the scalp are continuous analog signals that need to be converted into digital representations for the computer to process them. This is accomplished by using an analog-to-digital converter (ADC) which samples (i.e., records) the continuous analog signal at discrete time intervals. The faster the sampling rate (i.e., the larger the number of samples recorded in a unit time [usually seconds]), the better the signal representation in the digital form. The sampling rate for any signal should be calculated from the Nyquist criterion[8]. These signals are suitably preprocessed and then fed to a computer which processes them and uses them in order to accomplish the desired tasks. The preprocessing normally includes a notch filter of 50 Hz to filter the power line interference as well as a band-pass filter of 0.1−100 Hz to filter the movement artifacts, which are potentials generated by blinks, muscle movement and cardiac activity. The EEG signals of interest are typically between 2 and 60 Hz.

The EEG signal is acquired at a specific sampling frequency by using two or more (sometimes 64) channels (as per the 10−20 electrode system [cf. Figure 2.5]). The EEG in the work discussed in this text was acquired using a g.MOBIlab+ (multimodal biosignal data acquisition on a standard PC) and g.USBamp (24 Bit biosignal amplification unit) from g.tec Medical Engineering GmbH, Austria [96] shown in Figure 2.8.

g.MOBIlab+ and g.USBamp are g.tec'z' [98] (which can be in the form of a visual flash or an auditory sound). The P300 is one of the most common ERP responses used within the BCI technology, beginning with the works of Farwell and Donchin et al. [99,100]. The P300 wave [5,101] is evoked in the EEG as a positive peak at about 300 ms elicited by the desired response after a stimulus. The type of stimulus does not matter, i.e., it may be visual or auditory. In simple terms, the P300 is a measured brain response that is the direct result of any thought or perception. VEPs are also ERPs which are generated from the brain's primary visual cortex in response to a visual stimulus such

---

[7]CMRR is defined as the ratio between the differential-mode gain and the common-mode gain. The noise from the environment is generally called the common-mode gain and may include power line signals, inadequate grounding as well as power supply leakage. Thus, a higher CMRR is preferred as a means of reducing the effects of these unwanted signals.

[8]The sampling frequency should be at least twice the highest frequency contained in the signal. Since the EEG signals contain information in the 30 Hz range, they should be sampled at least at 60 Hz (2×30 Hz). The usual practice is to sample the EEG at a sampling frequency of 250 Hz.

FIGURE 2.8   gUSBamp and gMobilab⁺. *Reproduced with permission from g.tec medical engineering GmbH Copyright 2012 [96].*

as a flashing light [102,103]. The stimulus is presented repeatedly at a certain frequency while averaging the time locked response to several identical stimuli. Thus, the VEP is obtained by averaging the response of several sequential stimuli. The VEP is characterized by a negative peak at around 100 ms which is later followed by a positive peak at around 200 ms. These potentials are being used as clues to indicate the direction of the user's gaze [18,20,104]. The SSVEP is also a natural response to visual stimulations at specific frequencies. When excited by a visual stimulus ranging from 3.5 Hz to 75 Hz, the brain generates an electrical activity at the same or a harmonic of the frequency of the visual stimulus. Hence, these signals are used to understand which stimulus the subject is looking at, by showing them stimuli with different flashing frequencies [105,106]. The mVEP is a scalp potential of visual motion response and reflects the activation of the dorsal visual stream and the magnocellular pathway [107,108]. The mVEPs elicited by an abrupt visual motion onset of a structure display a dominant motion-specific negative peak at about 160 ms after the motion onset [109]. The mVEP is promising for BCI applications due to its large amplitude and low inter- and intra-subject variabilities [110]. The SCP electrical activities reflect the slow potential variations in the cortex and normally last for about 0.5–10 s [19]. The amplitude of these potential shifts usually varies within a range of 10 to 100 μV root mean square (rms), and reaches a maximum at the vertex (position of Cz electrode as per the international 10/20 system) [111]. Positive SCPs are associated with reduced cortical activation and negative SCPs are generally related to movement. Birbaumer and his colleagues demonstrated that adequately trained people can in fact control these potentials [112]. SCPs have been investigated in many BCI applications [113–115].

2.4.1.1 SMR BCIs

MI-based BCI is the focus of this book, primarily because it is a more intuitive way of expressing a desire towards a specific task of interest.

In MI-based BCIs, the intention of the subject is translated into a control signal by classifying the specific EEG pattern which is characteristic of the imagined task, e.g., movement of the hand and/or foot. These event-related phenomena represent frequency-specific changes of the ongoing EEG activity and may consist, in general terms, of either decreases or increases of power in given frequency bands [116], typically say in the μ (8−12 Hz) and β (13−30 Hz) frequency bands. This may be considered to be due to a decrease or an increase in the synchronization of the underlying neuronal populations, respectively. The former case is called event-related desynchronization [117,118] and the latter event-related synchronization (ERS) [119]. As discussed previously, it is possible to classify MI-related EEG signals based on μ and β rhythms which originate in the sensorimotor region [120]. Some β rhythms are independent EEG features, while some are harmonics of μ rhythms [19]. The μ rhythms display ERD over the contra-lateral hemisphere [121] which is normally accompanied by a β ERS [122] over the contra-lateral sensorimotor hemisphere. Therefore it is possible to classify μ and β rhythms [120] to achieve independent control, so the majority of BCI research focuses on these two bands [65]. Several research groups work on SMR BCIs [78,114,123−130]. A recent observation made in reference [131] suggests significantly higher activation of the supplementary motor areas (SMA)[9] for MI and motor observation tasks in high-aptitude users. This paper also demonstrates that acquisition of a sensorimotor program reflected in SMR-BCI-control is tightly related to the recall of such sensorimotor programs during observation of movements and unrelated to the actual execution of these movement sequences. It is important to note here that BCIs based on SMR [132] and SCPs [133] require no external stimuli, but do require user training, while the BCIs based on SSVEP [134,135] and on the P300 [100] component can be set up with almost no training, but they do require external stimuli. Another major difference is that the BCI users rest in between the trials in the MI-based techniques, whereas they do not have any time for rest in the latter techniques.

The raw EEG has a very low signal-to-noise ratio (SNR) and artifacts are present, due to interference from the mains, impedance fluctuations due to loose placement of electrodes, minor body movements/motion artifacts leading to electrode movements over the skin, electromyogram (EMG)/electrooculogram (EOG)/electrocardiogram (ECG) interference, eye blinks, other brain activity, breathing, biting, sweating, noise

---

[9]The supplementary motor area (SMA) is a part of the primate cerebral cortex and has long been thought to have a special role in the internal generation of complex movements [374], i.e., involved in planning complex movements and in coordinating movements.

introduced due to instrumentation or electronic devices, etc. Thus various forms of preprocessing are carried out to remove such unwanted components from the EEG signal, otherwise they may bias the analysis of the EEG and lead to wrong conclusions [136,137]. Preprocessing thus forms a very important part of the BCI process and if carried out properly leads to cleaner EEG signals. The next section details the importance of the preprocessing approach in EEG signal processing.

2.5 PREPROCESSING: A SIGNAL ENHANCEMENT REQUIREMENT ALONG WITH NOISE REDUCTION

This section describes the signal preprocessing requirement and common methodologies that have been implemented by various research groups. The primary requirement of the preprocessing stage in BCI is to filter out the noise, artifacts or unwanted signals that are embedded within the EEG. In view of the fact that EEG-based MI classification depends solely on the EEG as the signal of interest, the signals generated by muscle movement (known as electromyogram [EMG]), the activity due to movement of the eyes (known as electrooculogram [EOG]) and the electrocardiogram (ECG) should also be considered as unwanted signals or artifacts. The EMG activity is the most common artifact, and exists in the frequency range of 20 and 200 Hz whereas movement artifacts appear at frequencies of less than 10 Hz [138]. Similarly, the EOG exists within the frequency range between 0.1 and 38 Hz, but typically below 20 Hz [139]. The ECG artifacts are related to the field of the heart potentials over the surface of the scalp. These artifacts are located in the lower frequency ranges. There are various techniques, discussed below, to remove artifacts [136,140,141]. Most of the BCI groups utilize a band-pass filter as one of the most common preprocessing tools to extract information related to the EEG frequencies of interest. There is minimum probability of artifacts in the region of 8−12 Hz (lower frequency band) and 16−24 Hz (higher frequency band) because the other frequencies have been removed by way of band-pass filtering. These optimal bands of interest may vary from one subject to another, and hence finding subject-dependent frequency bands is necessary to maximize BCI performance. A notch filter of 50 Hz can be utilized to remove the common electrical power line interference/artifact. A good preprocessing tool can enhance the performance of the complete BCI [142,143] while using the same feature extraction (FE) and classification processes, so preprocessing the EEG signal is often considered the most important BCI stage. A few of the best-known preprocessing techniques commonly employed in BCI systems are therefore discussed in the following sub-sections.

2.5.1 Referencing Method

Referencing methods are considered as spatial filters. The primary assumption with these methods is based on the principle that the activity over the whole or complete head (skull) at every moment sums up to zero. Two frequently used referencing methods in BCI are the common average reference (CAR) and the Laplacian method. The CAR is a computationally inexpensive background-noise-removing method, which re-references the signal to a common average across all the electrodes by subtracting from each sample the average value of the signal across all the electrodes. Therefore, the average of all the activity represents an estimate of the activity at the reference site and then subtracting this average produces in principle a de-referenced solution.

CAR is mathematically expressed as:

$$v_i' = v_i - \frac{\sum\limits_{j=1}^{n} v_j}{n} \tag{2.1}$$

where v_i is the potential difference between the ith electrode and the reference and n is the total number of electrodes in the montage. It is important to note here that when $i = j$, the CAR can over-fit because the ith electrode potential is reduced due to the averaging process. CAR over-fitting can be addressed by not including the ith electrode for the summation $(\sum_{j=1}^{n} v_j)$ in (2.1). CAR has been widely used in BCI systems where the noise and artifacts are evenly distributed over the EEG montage [144,145].

Contrary to this, the Surface Laplacian method at a specific electrode position is derived by subtracting a combination of the signals from a set of surrounding electrodes from the central electrode signal. It is mathematically computed as:

$$v_i^L = v_i - \sum\limits_{j \in N_b} g_{ij} v_j \tag{2.2}$$

where:

$$g_{ij} = 1/d_{ij} \sum\limits_{j \in N_b} d_{ij} \tag{2.3}$$

Here, N_b is the number of electrodes surrounding the ith electrode and d_{ij} is the distance between the ith and the jth electrodes. These neighboring electrodes are normally tuned by selecting the explicit electrodes that maximally accentuate the relevant signal and maximally attenuate other non-relevant EEG activity and non-EEG artifacts. This approach is discussed by McFarland et al. [145] and highlights the usefulness of the same in BCI referencing and artifact suppression.

Both the CAR and Laplacian methods have shown better results over the earlobe reference methods (i.e., an electrode such as the one near an earlobe is the reference) with the BCI systems by using the μ and β rhythms. The primary reason for this is that both these approaches employ high-pass spatial filters which enhance the local sources (i.e., the μ and β rhythms) while simultaneously reducing the distant sources (e.g., EMG, eye movements, etc.) [145].

2.5.2 Principal Component Analysis [PCA]

Principal Component Analysis (PCA) [146] is a common statistical technique for identifying and re-referencing the data by linear mapping, which transforms a number of possibly correlated variables into a smaller number of uncorrelated variables known as principal components. The primary axis or the first principal component is calculated such that it accounts for the largest amount of variability in the data. Similarly, the subsequent components or axes are calculated and they account for the direction of the remaining variability but in decreasing order of the amount of variability in the data. The subsequent axes thus represent the direction of the next largest variation and so on. Since the transformed data have most of the variation in the first few components, the remaining components can be ignored to reduce the dimensionality but simultaneously not significantly compromise the accuracy of the data representation [147].

Applying PCA to a set of data is a five-step process. The steps involved are the following:

1. The first step is to subtract the mean from each of the data dimensions. This generates a data set whose mean is zero, i.e., the data matrix is centered.
2. The next step is to calculate the covariance matrix of the centered data.
3. The next task is to calculate the eigenvectors and the eigenvalues of the covariance matrix. An eigenvalue represents the amount of variance within a given component.
4. These eigenvectors are then ordered according to their eigenvalues, highest to lowest. Thus, these are the principal components of the data set. This is necessary to order the components in the order of significance. The final data set will have fewer dimensions than the original if the components with lesser significance are removed.
5. The final step is to form a feature vector by taking the retained eigenvectors. The eigenvectors in the feature vector should be arranged column wise.

$$\text{Feature Vector} = (ev_1 \; ev_2 \; .. \; ev_n)$$

The *FeatureVector* is then transposed and the transposed mean is subtracted from the original data i.e., the centered data, to obtain the PCA data. Thus, the data in its final form is represented as:

PCA_data = Transposed_FeatureVector

× Transposed_OriginalMeanSubtractedData.

Hence, if the data were initially n dimensional, if the first p eigenvectors are chosen, the final data set will consist of only p dimensions.

2.5.3 Independent Component Analysis [ICA]

The phrase 'Independent Component Analysis' first appeared in reference [148]. Independent Component Analysis (ICA) is a statistical and computational technique for revealing the hidden sources/components that underlie sets of random variables, measurements or signals [149]. Work in ICA from a statistical point of view was first carried out by Comon [150] and later applied to EEG by Makeig et al. [151]. A common application of the ICA separation problem is to the so-called 'cocktail party' problem, where the voices of several people in conversation are recorded. The objective of the problem is to find one individual's voice signals from the mixed recording. A common assumption made when applying the ICA technique is that the unknown underlying sources are independent of each other and have been linearly combined to form a mixed signal. The ICA returns the independent components when this independence assumption is correct. This assumption leads to a linear ICA model.

The linear ICA model can be understood in its simplest form where X is modeled as:

$$X = AS \tag{2.4}$$

Here, the source components $\{S_j\}$ are assumed to be statistically independent, and A is a mixing matrix with unknown coefficients that is responsible for the component separation. The source component signal $\{S_j\}$ at a time instant t is mixed to form the observed signal X. The task of ICA is to estimate the mixing matrix A and recover the source component S. The dimension of the mixing matrix A is mXn, where m indicates the number of mixed signals and n indicates the number of sources. Generally, the number of sources is less than or equal to the number of mixed signals.

In a nonlinear ICA model, the assumption made is that the mixing of the sources may not be linear. The solution of the nonlinear ICA problem is usually highly non-unique and the indeterminacies in the separating solutions are much more severe than in the linear case and hence nonlinear ICA is difficult to apply [152,153].

The recording of the EEG signals is assumed to be instantaneously and linearly mixed by the underlying brain sources [154,155]. Therefore, the noise added to the observation can be assumed to be special underlying 'sources' participating in the mixture. Hence, most ICA research is based on the simplified ICA model [154]. ICA is now widely used in BCI research for preprocessing the acquired EEG signals [140,156,157].

PCA extracts signals that are uncorrelated, while ICA extracts signals that are statistically independent (or are from independent source signals). This is the most relevant distinction between these approaches.

2.5.4 Common Spatial Patterns [CSP]

Common Spatial Patterns (CSP) is a discriminative user-specific filtering technique for signal enhancement. It detects patterns within the complex EEG signals by incorporating the spatial information of the EEG [97]. This is implemented by utilizing the covariance matrices as the basis and seeking a discriminative subspace such that the variance for one class is maximized and the variance of the other class is minimized at the same time. Given N channels of EEG for each trial X of class 1 and class 2, the CSP method gives an $N \times N$ projection matrix [141,158−160]. This matrix is a set of subject-specific spatial patterns that reflect the specific activation of cortical areas during hand movement imagination [161]. The decomposition of a trial X is:

$$Z = WX \tag{2.5}$$

where W is the projection matrix. The mapping projects the variance of X onto the rows of Z, resulting in a new time series such that the variance for class 1 is largest in the first row of Z and decreases with increasing number of the subsequent rows.

The CSP approach does not require any *a priori* selection of subject-specific frequency bands, but at the same time, if this spatial filter is applied to any randomly selected frequency band, then the performance of the CSP is poor [162−164]. Therefore, selecting a frequency band over a wide range for any subject is a common practice for using the CSP algorithm [165], and gives better classification than the bands over a narrow range [159]. The knowledge of the frequency bands becomes necessary later on for the band power and frequency-estimation methods [141]. The CSP approach detects the spatial patterns in the EEG to achieve improvements in the accuracy of classification, and hence even a minor change in the position of the electrodes may lead to a loss in the gained improvements. Therefore, care must be taken in placing the electrodes for all the trials and sessions [160]. Different variants of CSP filtering methods have emerged in the recent past, for better performance. Some of these are Common Spatio-Spectral Pattern (CSSP) [163],

Common Sparse Spectral Spatial Pattern (CSSSP) [163,166], Sub-Band CSP (SBCSP) [167], Filter-Bank CSP (FBCSP) [162] and Regularized CSP (RCSP) [164]. Various research groups have effectively used CSP in BCI [168,169].

2.5.5 Neural time series prediction preprocessing [NTSPP]

The neural time series prediction (NTSPP) approach, shown in Figure 2.9 and introduced in reference [39], is a multistage signal processing framework which permits multiple-step-ahead prediction of the EEG time series. Here, different prediction models (PMs) are trained to specialize in predicting different EEG signals. The basic purpose of these PMs is to exploit the differences in predictions by different specialized predictor networks for predicting different EEG signals. The NTSPP has been shown to increase feature separability by mapping the original EEG signals via time series-prediction onto a higher dimensional space [46,169−172].

The work in reference [169] shows that NTSPP has the potential to enhance the information from the EEG with the existing BCI methods using less channels. Here, non-subject-specific spectral filters are employed in conjunction with the CSP approach and tested with different classifiers.

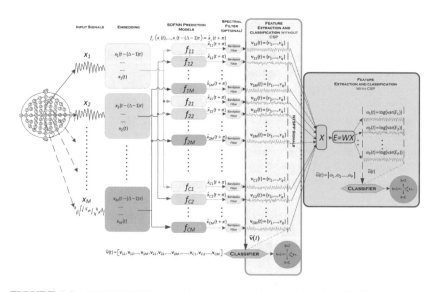

FIGURE 2.9   NTSPP-CSP framework. *Reproduced with permission from [172].*

The two channels C3 and C4 used for data acquisition produce four time series (two for each of the two classes — left (C3, C4) and right (C3-C4)) as shown in Figure 2.9. Using the NTSPP framework, different PMs are used to predict one step ahead from the data, i.e.:

$$\text{no. of predicted time series} = \text{no. of EEG channels} \times \text{no. of classes} \qquad (2.6)$$

For the prediction of a sample at time $t + \pi$, the EEG signal from time t to $t - (\Delta - 1)\tau$ is used. Here parameter Δ is the embedding dimension and:

$$\hat{x}_{ci}(t + \pi) = f_{ci}\langle x_i(t), \ldots, x_i(t - (\Delta - 1)\tau\rangle \qquad (2.7)$$

where:

τ is the time-delay constant,
f_{ci} represents the prediction model trained on the ith EEG channel,
π is the prediction horizon.

2.5.6 Kalman Filter

In 1960, R.E. Kalman published his famous paper describing a recursive solution to a linear filtering problem [173]. The Kalman filter (KF) is a two-step approach of prediction and correction, and infers the parameters of interest from indirect, inaccurate and uncertain observations [174]. In the prediction stage, the state is predicted using a dynamic model, while in the second stage, the predicted state is corrected using an observation model such that the error covariance of the estimator is minimized. Being recursive, new measurements can be processed as they arrive. A good introduction to the Kalman filter is described in reference [175], and it is explained in Figure 2.10. In KF, a model in which noise is uncorrelated in time describes the system; i.e., if the value of the noise is known now this knowledge does not help in predicting the noise values at any other future time. The development of the KF under the restrictions of linearity results in the state of the system being determined by its mean and covariance matrices. KF has been investigated in EEG-BCI [35,176,177] as it is expected to handle noisy brain signals effectively. However, the application of the KF to nonlinear systems can be difficult, and therefore a modification in the form of extended KF (EKF) [178] was devised for non-linear systems by Stanley F. Schmidt. EKF simply linearizes all nonlinear models so that the traditional linear KF can be applied [179]. Here the state equations are repeatedly linearized about each estimate once it has been computed through first order approximations. Therefore, EKF may result in large errors and sometimes divergence of the filter [178]. EKF has been investigated in BCI applications in reference [35].

Kalman filter tries to estimate the state x of a discrete-time controlled process governed by a linear stochastic difference equation $x_k = Ax_{k-1} + Bu_{k-1} + w_{k-1}$.

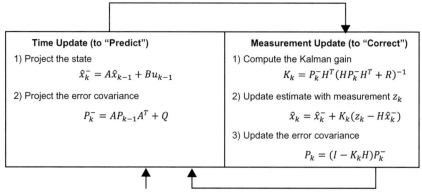

Time Update (to "Predict")	Measurement Update (to "Correct")
1) Project the state $$\hat{x}_k^- = A\hat{x}_{k-1} + Bu_{k-1}$$ 2) Project the error covariance $$P_k^- = AP_{k-1}A^T + Q$$	1) Compute the Kalman gain $$K_k = P_k^- H^T (HP_k^- H^T + R)^{-1}$$ 2) Update estimate with measurement z_k $$\hat{x}_k = \hat{x}_k^- + K_k(z_k - H\hat{x}_k^-)$$ 3) Update the error covariance $$P_k = (I - K_k H)P_k^-$$

Initial estimate for $\hat{x}_{k-1}$ and P_{k-1}

where

w_k represents the process noise, Q is the process noise covariance and R is the measurement noise covariance. The $n \times n$ matrix A relates the state at the previous time step $k-1$ to the state at the current step k. The $n \times 1$ matrix B related the control input u to the state x. The $m \times n$ matrix H relates the state to the measurement Z_k. $P_{\bar{k}}$ represents the *a priori* estimate error covariance and P_k represents the *a posteriori* estimate error covariance. The $n \times m$ matrix K is the gain that minimizes P_k.

FIGURE 2.10   Operation of the Kalman filter (to 'predict' and 'correct').

2.5.7 Autoregressive (AR) Modeling

The AR model is a linear predictive modeling technique. It attempts to predict the signal sample based on previous signal samples by using the AR parameters as coefficients. The number of samples used for prediction determines the order of the model, n_r.

$$x[k] = e[k] + \sum_{r=1}^{n_r} a_r x[k-r] \qquad (2.8)$$

Here, $e[k]$ is a white noise or residual error with zero mean and a_r is the filter coefficients. The AR parameters can be estimated using several techniques such as Kalman filter, Yule-Walker, Expectation-Maximization, least-square, Burg, forward-backward algorithm, etc. [49,180−183]. The combination of Kalman filtering in estimating AR parameters produces the methodology of Adaptive Autoregressive (AAR) modeling. EEG signals have been modeled for BCI by the AAR process whose parameters are extracted using the Kalman filter [184]. However, since the EEG signal is noisy and chaotic, the characterization process of features by AR/AAR models is difficult and can lead to very bad classification performances [38].

2.5.8 Summary

Most of the filtering algorithms discussed are developed on the assumption of *a priori* information or linearity or that the deterministic part of the nonlinear signal can be linearized about a nominal state trajectory. Again, these filters may also cause large errors and may sometimes diverge. Therefore, this book presents an information processing architecture using concepts from quantum mechanics that does not require any *a priori* information about the nature of the noise present in the signal. A combination of a layer of neurons in the spatial axis is treated as a probability density function (*pdf*) that recurrently evolves under the influence of the Schrodinger Wave Equation (SWE) and an appropriate set of learning rules. This architecture results in an efficient computational algorithm referred to as the recurrent quantum neural network algorithm (RQNN) and to some extent solves the complex problem under consideration (cf. Chapter 3 and Chapter 5).

2.6 FEATURE EXTRACTION

Feature extraction (FE) provides the description of the signal to be classified in terms of its invariant distinguishing characteristics [185]. The features to be extracted from the noisy signal should ideally be an enhanced representation of the signal and be invariant to noise. Various approaches have been investigated for producing high-quality, practical BCI systems using the electrophysiological information discussed above. This section presents some of the best known and most commonly used FE methodologies implemented by different research groups. Features such as the amplitude values of the *pdf* of the EEG, power of the EEG signal, Hjorth, spectral density, autoregressive (AR), time-frequency (t-f) features, etc. have been utilized by various research groups [37,48, 49,186,187]. This section begins with the discussion of band power features.

2.6.1 Band Power Features

Power in general is defined as the energy over a unit time interval. This classical approach of band power (BP) FE can be realized to quantify the ERS and ERD phenomena discussed in Section 2.4.2. This method is based on calculating the squared amplitude of the signal over a small time window. Since this technique is based on a windowing approach, it primarily assumes the signal to be stationary within that window. Therefore, this method should preferably be utilized with smaller window segments in order to reduce the inherent non-stationary effects present in the EEG within that window. This approach typically uses two frequency bands; the μ band (8−12 Hz) and the β band (13−30 Hz) for the purpose of FE, although, as discussed previously, the range of these frequency bands may vary from one subject to another. The fine tuning of the ranges of the different frequency bands also plays a crucial role in the success of the BCI based on this approach. Pfurtscheller et al. [122] used the BP features and demonstrated that for each subject, different frequency components in the μ and β bands provided best discrimination between the left and the right hand movement imagination. As this FE method is simple, easy to use and computationally efficient and extracts the most relevant information, it is one of the most commonly used approaches in MI-BCI [188]. A limitation of the BP technique is that it restricts attention to a particular frequency band. Another approach, based on Fourier methods, is to divide the epoch of EEG data into smaller segments and calculate the power spectral density (PSD) for each segment [189]. This is discussed in the next sub-section.

2.6.2 Power Spectral Density Features

The PSD approach is a different way of estimating band power. The PSD is known as a spectrum, and it describes how the power of a signal is distributed over a frequency range. The PSD features capture the frequency content of a stochastic process and are mathematically expressed as the Fourier transform of the signal's auto-correlation function [190]. The PSD approach for EEG FE is very popular [49,191−193]. The simple frequency transforms, such as the PSD, are normally able to bring out the overall frequency content of the signal, and this helps to discriminate between the different MI EEG signals. The PSD approach has been investigated extensively in reference [49] and has demonstrated superior robustness and consistency even across several classifiers.

A periodogram is another computationally efficient approach for the estimation of the power spectrum. This method divides the complete signal into smaller blocks of data, known as windows, and then computes the squared magnitude of the windowed Discrete Fourier Transform (DFT) of each of the blocks. The average of these blocks, which are sometimes overlapped, gives an estimate in the form of PSD. A better frequency resolution is obtained when a longer window is utilized, i.e., the size of an individual block is larger, but with this approach there are a smaller number of blocks to average and this may lead the estimated features to have higher variance. The size of an individual block (window size) and the amount of the overlap is normally chosen to suit the concerned application. The periodogram-based method has been implemented for detecting SSVEP potentials in reference [194].

Another common technique for estimating the PSD is by using the Yule-Walker AR method. This method is also known as the autocorrelation method and fits an AR model to the windowed input data. Thus, this approach determines the AR parameters that represent the EEG segments within the window. This approach has been implemented in many signal processing approaches in BCI [49,195].

2.6.3 Time-frequency Method

A signal, as a function of time, may be considered as a representation with perfect temporal resolution. Similarly, the magnitude of the Fourier transform (FT) of the signal may be considered as a representation with perfect spectral resolution but with no temporal information. However, a combination of time-frequency representation that maps a one-dimensional signal into a two-dimensional function of time and frequency can be used to analyze the time-varying spectral content of non-stationary signals [196] such as the EEG. The short-time FT (STFT) is widely applied in EEG analysis, despite the fact that the time and frequency resolutions are linked together and have to be compromised [197−199]. However, the wavelet transform allows for greater flexibility in this regard and is also robust to noise, hence most of the designs that employ time-frequency representation methods use wavelet-based feature extraction algorithms. A wavelet transform uses a variable window size over the length of the signal, which allows the wavelet to be stretched or compressed depending on the frequency of the signal [200]. This results in excellent feature extraction from non-stationary signals such as EEGs [201−207]. The continuous wavelet transform is defined as:

$$CWT(a, b) = \int_{-\infty}^{+\infty} x(t)\Psi_{a,b}^*(t)dt \tag{2.9}$$

where $x(t)$ represents the analyzed signal, a and b represent the scaling factor and translation along the time axis (shifting coefficient), respectively. $\Psi_{a,b}^*(\cdot)$ is obtained by scaling the wavelet at time b and scale a as:

$$\Psi_{a,b}^*(t) = \frac{1}{\sqrt{|a|}}\Psi\left(\frac{t - b}{a}\right) \tag{2.10}$$

Here, $\Psi(t)$ is a continuous function in both time and frequency domain referred to as the mother wavelet [208], and * represents the operation of complex conjugate. As a decreases the time resolution will be better but the frequency resolution will be poor [209]. The continuous wavelet transform is continuous in the sense that during analysis, the wavelet is shifted smoothly over the full domain of the function being transformed. However, calculating wavelet coefficients for every possible scale can represent a considerable effort and result in a vast amount of data. On the other hand, the discrete wavelet transform is a sampled version of the continuous wavelet transform, thereby making it computationally very efficient without compromising the accuracy. An important step in systems using wavelets is to select a subset of wavelet coefficients that are relevant for classification; i.e., selecting the regions in the time-frequency plane at which signals can be classified with high accuracy and can be achieved with the help of feature selection algorithms [210].

2.6.4 Hjorth Features

Hjorth features [53] are represented in the time domain and were introduced to give relevant information about the continuous EEG activity in the temporal domain through the trio of combinations of conventional time domain based descriptive statistics Hjorth parameters namely; activity, mobility and complexity. The Hjorth parameters can be employed as discriminative features to describe the signal in terms of its morphological characteristics.

The first Hjorth parameter is a measure of the mean power of the signal characteristics in terms of activity (variance of the signal) and is mathematically defined as:

$$Activity(y) = \sum_{i=1}^{N_s} \frac{(y(i) - a)^2}{N_s} \qquad (2.11)$$

The second Hjorth parameter, called mobility, is an estimate of the mean frequency [186] and is defined as:

$$Mobility = \sqrt{\frac{var(y')}{var(y)}} \qquad (2.12)$$

The third Hjorth parameter, called complexity, is an estimate of the bandwidth of the signal and is defined as:

$$Complexity = \frac{Mobility(y')}{Mobility(y)} \qquad (2.13)$$

For (2.11) to (2.13) above, y is the signal, y' is the derivative of the signal, N_s is the number of samples in the window and a is the mean of the signal within the window. The Hjorth parameters have been utilized as EEG features by many researchers working in the field of EEG-based BCI systems [36,170, 211−214].

2.6.5 Hilbert-Huang Transform

The Hilbert-Huang transform (HHT) is a way to decompose complicated data or a signal into fewer components, known as intrinsic mode functions (IMF), by using the empirical mode decomposition (EMD) [215] method, thereby obtaining instantaneous frequency data [216–218]. The EMD method breaks down the signal without leaving the time domain because the decomposition is based on the local characteristic time scale of the data. The HHT method is therefore applicable for data that are nonlinear and non-stationary. This method is intuitive, direct and adaptive, with an *a posteriori* defined basis from the decomposition method based on and derived from the data. The simple assumption here is that any data consist of different, simple, intrinsic modes of oscillation, and each intrinsic mode represents a simple oscillation, which will have the same number of extrema and zero-crossings. Furthermore, the oscillation will also be symmetric with respect to the 'local mean' [217]. This transform can be applied as a feature extraction approach to reveal better features from the raw EEG signals.

2.6.6 Summary

The goal of FE techniques is the first step in transforming the raw neurophysiological signals into a more suitable set that can subsequently be used in the next stage. A thorough comparison of different power FE techniques in the context of BCI is presented in references [49,219]. However, the results presented in reference [219] find a wavelet-based technique to be better suited, while reference [49] suggests PSD as a more robust technique for BCI. This difference between the two comparative studies is probably due to the difference in cross-validation of the wavelet parameters and keeping more frequency bands. However, it is also possible to combine features from different domains say, the temporal, the frequency or the spatial domains. Combining features from different domains can potentially increase the Classification Accuracy (CA) by being more separable than the individual set for, e.g., combining temporal and frequency domain features [220]. Therefore, the work presented in this book has combined the Hjorth parameters and the simple bandpower features for the discrimination of mental tasks using the RQNN filtered EEG signals (cf. Chapter 5). This is because the calculation of Hjorth parameters is based on variance and the computational cost of this method is considered low compared to the other methods. Thus, the calculation of Hjorth parameters is simple and efficient. In addition, the time domain orientation of Hjorth representation may prove suitable for situations where ongoing EEG analysis is required [186]. The next section discusses the classification process.

2.7 CLASSIFICATION

Classification methods are employed to recognize the patterns of EEG activities, i.e., to learn the mapping between the EEG data and classes corresponding to mental tasks such as movement of the left hand [221]. Thus, the classifier lets us know the class or group to which the input signal belongs. There are various classifiers used in the BCI community, which can roughly be grouped into linear and nonlinear classifiers. Linear classifiers are the most commonly used in BCI, primarily due to their robust design and the fewer parameters that need to be tuned in comparison to nonlinear classifiers [222]. Linear classifiers decide the class membership of a sample depending upon a linear combination of the features; i.e., a linear function. For a two-dimensional case, the linear classifier is simply a line. If a nonlinear function is employed to separate the classes, it is said to be a nonlinear classifier. A comprehensive survey of different classifiers for BCI can be found in references [221,223]. Some of the major classifiers used in BCI classification include linear discriminant analysis (LDA), support vector machines (SVM), neural networks (NN), nonlinear Bayesian classifiers and nearest neighbor classifiers, and these are discussed in reference [221]. It was shown in the BCI Competition 2003 and 2005 that LDA performed as well as (and sometimes even outperformed) SVMs [224], and almost all the winning classifiers were linear [225,226]. It has been found that simple linear classifiers were just marginally worse than complex nonlinear methods [222]. It should be mentioned here that the presence of excessive noise and unwanted information can also lead to the failure of linear classifiers. Some of the most popular classifiers within the BCI community are discussed next.

2.7.1 Linear Discriminant Analysis Classifier

Linear Discriminant Analysis (LDA) is one of the most successful linear classifiers in BCI [188,221]. LDA produces a separating hyperplane described as $f(X) = W^T X + B$ such that X belongs to class 1 if $f(X) > 0$, and it belongs to the class 2 if $f(X) < 0$. This function is called the linear discriminant function and the decision boundary is the hyperplane with a set of points satisfying $f(X) = 0$. The mean of the data set belonging to each of the two classes is represented as μ_1 and μ_2, and the mean of the entire data set of the combined classes as μ_3. With an assumption of an *a priori* probability of the individual class as p_1 and p_2 respectively (this can numerically be assumed to be 0.5), μ_3 can be calculated as:

$$\mu_3 = p_1 * \mu_1 + p_2 * \mu_2 \qquad (2.14)$$

The linear discriminant function is a projection onto the one-dimensional subspace such that the classes would be separated the most. The basic idea behind Fisher's LDA[10] is to have a 1-D projection that maximizes the variance of the projected centers, while minimizing the variance of the projected data points of each class separately; i.e., maximizing the ratio of between-class variance to within-class variance. There is only one direction of the discriminating line that will yield the best separation results. One has to determine the coefficients of this discriminating function. LDA does not do any kind of transformation of the data but provides a better class separability by creating a decision boundary between the given classes. Two modified versions of the LDA are available in the form of Fisher's LDA (FLDA) and Stepwise LDA (SWLDA).

FLDA overcomes the shortcomings of LDA [227] primarily because it measures the separability of the classes by two criteria; firstly using the distance between the projected means of the two classes and secondly utilizing the extent of the variance of the projected data. For a single parameter, Fisher's ratio, F, is a measure of discriminability:

$$F = \frac{\text{Variance of the class means}}{\text{Average variance within a class}} \tag{2.15}$$

Fisher's criterion thus calculates a ratio of the between-class separability and the within-class variation. Larger Fisher criterion values demonstrate improved separation of two classes [228]. FLDA has been utilized for classification in BCI [229], and is preferred by many researchers.

The SWLDA [230] is an extension of the FLDA in which only those features that are suitable for the classification purposes are selected for the discrimination analysis [231,232]. A combination of forward and backward stepwise analysis is implemented, wherein the input features are weighted using ordinary least squares regression (equivalent to FLDA) to predict the target class labels. Starting with no initial features in the discriminant function, the most statistically significant input features for predicting the target label are added to the discriminant function. After each new entry to the discriminant function, a backward stepwise analysis is performed to remove the least significant input features. This process is repeated until the discriminant function includes a predetermined number of features, or until no additional features

---

[10]The terms Fisher's linear discriminant and LDA are often used interchangeably, although Fisher's original article (1936) does not make the assumption that a LDA makes in the form of normally distributed classes or equal class covariance.

satisfy the entry/removal criteria [232]. Thus, this approach reduces the number of features required for classification and in the process also reduces the computational expense.

Overall, the LDA is robust, easy to use with lower complexity, has very low computational requirements and generally provides good results. It has been utilized within the BCI community by many research groups, some of which are discussed in references [6,22,44,81,187, 233–238].

2.7.2 Support Vector Machine Classifier

The Support Vector Machine (SVM) classification method was introduced in 1992 by Boser, Guyon and Vapnik in reference [239]. The idea (in SVM) is to find an optimal hyperplane that separates the feature points of the two different classes by the largest possible margin in the feature space. This can be achieved either within the same feature space, or by projecting the features into a higher dimensional plane where classification is easy. The points X, which lie on the hyperplane, satisfy $W^TX + B = 0$, where W is normal to the hyperplane, $\frac{|b|}{\|W\|}$ is the perpendicular distance from the hyperplane to the origin (i.e., bias $B = 0$), and $\|W\|$ is the Euclidean norm of W. The task then is to identify the feature samples that are closest to the decision boundary. These feature samples determine the 'margin' by which the two classes can be separated (cf. Figure 2.11). This margin is defined as $d_+(d_-)$; i.e., the shortest distance from the separating hyperplane to the closest positive or negative value of the data. The objective is to maximize this margin such that it contains all the points but with a very small error, i.e., there is preferably no overlap of class 1 samples misclassified as class 2 samples. The points for which the equality for $W^TX + B \geq 1$ is true lie on the hyperplane H1: $W^TX + B = 1$ with normal W and perpendicular distance from the origin being $\frac{|1-b|}{\|W\|}$. Similarly, the points for which the equality

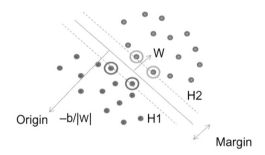

FIGURE 2.11   Separating hyper-planes for SVM (support vectors encircled).

for $W^TX + B \leq -1$ is true lie on the hyperplane H2: $W^TX + B = -1$ with normal W and perpendicular distance from the origin being $\frac{|-1-b|}{\|W\|}$. Thus the total margin is $\frac{|2|}{\|W\|}$ with $d_+ = d_- = \frac{|1|}{\|W\|}$.

2.7.3 Regression Classifier

Regression classifiers are based on logistic regression [240] and are one of the best probabilistic classifiers. The most common form of regression is to fit a straight line through a set of data points. The values of the data points are assumed to come from a normally distributed random variable with a mean that can be obtained as a linear function of the predictors and with a constant variance. To classify a data point x as belonging to a class n, let the probability that the data point x belongs to the class corresponding to β be:

$$\rho(x) = \frac{1}{1 + e^{-\beta x}} \tag{2.16}$$

Thus, there will be one β for each class. An iteratively reweighted least squares (IRLS) method[11] [241] is normally used to obtain β. The algorithm needs to be run once for each class to obtain, as discussed, one β for each class. The class with the highest value of β is chosen as the class to which the data point belongs. The basic difference between SVM and regression classifiers is that SVMs are typical learning machines based on clustering, classifying and ranking the data. However, the backbone of regression classifiers is the predictive analytics. Typically, these classifiers fail when there are more than two segments and hence a number of logistical functions should be combined to approach this issue. Neural networks can be considered as an implementation of this idea, as they combine several perceptrons to solve a complex classification process [240]. Regression classifiers have been used for EEG signal classification [242].

2.7.4 Summary

Classifiers play an important role in both the offline and the online stages. During the offline stage, the parameters of the classifier are tuned by using the training dataset belonging to the input signal. The classifier parameters thus obtained are kept fixed for use during the online process. Adaptive classifiers based on several approaches

---

[11]The method of IRLS is used to find the maximum likelihood estimates of a general linear model by an iterative method.

such as updating the mean as well as covariance of the class distributions continuously in time [237,238,243], investigating the shifts in data
distributions during the offline and online stages as well as in between
sessions [44], changes in EEG features during online sessions [44],
adapting the classifier parameters based on detection of error potential
that act as a reinforcement signal [244] are being investigated. The classification stage in most present-day BCIs is followed by a post-
processing and thresholding block. The classifier outcome in the form of
a time-varying signed distance (TSD), shown in Figure 2.12 (single trial),
is post-processed to take into account the highly varying/momentary
spontaneous TSD peaks, error removal [82], classifier bias removal [245]
or generating compatible control signals to operate a device. As discussed earlier, in accordance with the trial time (in Figure 2.12 it is 8 s),
the computer generates a beep signal that indicates the beginning of a
new trial, which typically starts at second 2. The user's choice has to be
registered within the specified time duration; which again is typically
the time of MI from 4 to 7 s. If the classifier output/TSD (cf. Figure 2.12)
during the MI period exceeds an upper threshold (which can be chosen
as any suitable value, say +0.75 in figure) and remains there for a substantial period of time, then the complete trial is designated to belong to
class 1. This substantial period depends upon what percentage of classifier outputs are desired above the threshold value to choose a winner
amongst the two classes for that particular trial. This substantial period

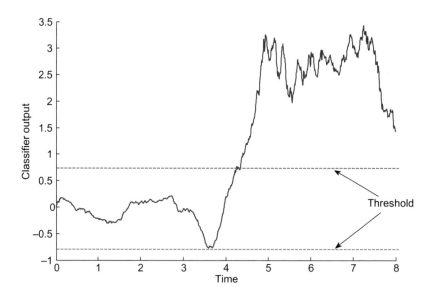

FIGURE 2.12   Understanding TSD through a single trial.

is also referred to as D in Fitts' law[12] [246]. Since the TSD is higher for a substantial period in the shown diagram, the class +1 is selected as the winning class. If the TSD had remained below the lower threshold (say, −0.75 in Figure 2.12) for a substantial period of time, then the trial is said to belong to class 2. If the continuous classification output had remained between the threshold values (for a substantial period, i.e., less than D), then it indicates a NC state from the BCI user. The thresholds can be chosen to be any suitable value. However, since the threshold is used here to produce a region in which NC is to be detected, a higher value of the threshold will result in more NC. Hence, care needs to be taken in setting the appropriate time threshold (using the Fitts' concept discussed in Section 2.12) and the distance threshold (which depends on the strength of the TSD) to conclusively ascertain the output class or the intention of the BCI user within the MI period. The third output state (i.e., the NC state) should thus be managed appropriately. The TSD may, however, not always be as clear as the one shown in Figure 2.12. Most of the time, the TSDs give ambiguous outcomes which may not be suitable for the smooth controlling of a device. Operation of the device requires a reliable, stable control signal and thresholding methodology. This has been investigated in the work presented in this book and has helped to improve the identification/ authentication capacity of the complete BCI system.

The next section discusses post-processing.

2.8 POST-PROCESSING

The basic purpose of post-processing is to appropriately enhance the TSD obtained from the classifier stage by making it more reliable and compatible for an end application and thereby improve the overall performance of the BCI. Post-processing filters out the spontaneous peaks in the classifier output (or the TSD). It is also a way to identify ambiguities from the classification stage. The post-processing approach carried out in the present work is primarily concerned with the latter issue; i.e., it is done to reduce the effect of momentary/spontaneous peaks in the TSD or the uncertainty in the classifier output by utilizing a dual-threshold windowing concept as explained in Section 2.8.2 below. Some of the common post-processing techniques used in BCI are discussed next.

---

[12]Fitts' law is used to model the act of pointing. D is the distance from the starting point to the center of the target while performing the act of pointing. In the present case, this refers to the task of performing the MI so that the TSD remains above the threshold. More about Fitt's approach is discussed in Section 2.12.

2.8.1 Confidence Intervals and Rejection

A common technique in BCI is to introduce the concept of thresholding and confidence intervals to reject ambiguous classifier outcomes and thereby avoid false activations. In such cases, the output of the system is only activated when the probability of an output being in an active state is greater than a pre-specified confidence threshold. A number of post-processing methods, including removing the bias of the classifiers, smoothing the outputs to remove jitters and applying intelligent thresholding to the classification outputs, are discussed in references [82,247].

2.8.2 Multiple Thresholding with Windowing Concept

The classifier output in the form of a TSD is normally jittery and contains transient peaks, as shown in Figure 2.13. If the TSD is above the positive threshold, the signal is considered to belong to class 1 and if the TSD is below the negative threshold then the signal is considered to belong to class 2. And if the TSD lies between these two thresholds, the signal is assumed to be in a NC state. The typical TSD shown in Figure 2.13 is for a single trial of duration 5 s in which the user performed the MI between second 1 and second 4. However, it seems that the trial belongs to class 1 around the region of 3.8 s while it belongs to class 2 around the region of 2.5–3 s and 3.9 s during the same single trial. Again, the trial seems to be in an NC state between the time

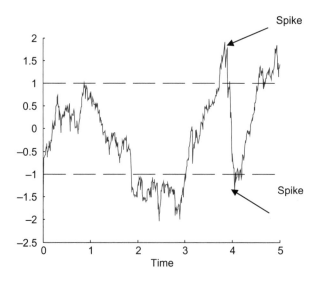

FIGURE 2.13   Multiple thresholds for a representative trial of 5 s duration.

interval 3 to 3.8 s. This may lead to contradictory information being transferred from the classifier stage to the device being commanded. A probable solution for rejecting these disputed and sometimes transient outputs is to use multiple thresholds in the form of 'how much' and 'how many.' This concept of multiple thresholds is a way of keeping a check on the classifier outputs within a specific timed window interval. The first is the normal threshold concept (cf. Figure 2.13). A 1 s window is assumed with the down-sampled TSD having 10 samples within this 1 second window. The multiple threshold concepts lie in counting the number of samples classified that belong to either of the two classes as well as the no-control state. The approach taken in the present work confirms that the number of these samples belonging to a specific class should exceed an index of difficulty level of say 70% (or seven samples within a 1 s window in the present case). This is primarily a necessity because, as seen in the TSD of Figure 2.13, there is a minor spike between second 3 and second 4 leading to a conclusion that the trial may belong to class 1. However, a closer investigation of the TSD at second 4 suggests that the trial could also have belonged to the class 2. Therefore, if an index of difficulty level is taken into consideration then it can avoid such debatable decisions. Thus, the population of samples belonging to a particular class should not only exceed its counterpart but should also exceed the index of difficulty level (which can be set as per the TSD) otherwise, the trial is considered to be in the no-control state. This helps to check false activations.

2.8.3 De-biasing

De-biasing the classifier output in its simplest form can be performed in an unsupervised manner by removing the mean calculated from a window of the recent classifier outputs from the instantaneous value of the classifier [82]. De-biasing can improve the online feedback response, however it may only provide a slight performance improvement [248] and therefore de-biasing is not much prevalent in the BCI system.

2.8.4 Error Potential (ErrP)

Error potential (ErrP) is the reaction of the BCI subject in response to an erroneous answer, i.e., a signal is generated by the brain in response to some error that has been made by the BCI user. Since this signal is indicative of the error made by the user, it can be utilized as a feature to signify whether the classifier outcome given by the classifier process at the end of a particular trial is correct or not; i.e., to confirm whether an outcome of a trial is as anticipated by the BCI user or not [245].

Thus, ErrP is looked upon as a post-processing block in many of the present day BCI systems [249–251]. It can also be utilized as a feedback element and has been shown to enhance the overall performance of a practical BCI robotic system, as discussed in [102].

With an understanding of the post-processing section, the signal processing aspects of the present day BCI system are said to be complete. Some prominent optimization techniques used for tuning various system parameters that can enhance the classification accuracy and the overall performance of the BCI are discussed next.

2.9 VALIDATION AND OPTIMIZATION TECHNIQUES

All of the approaches presented have parameters that should be set to their optimum values. Parameters are normally tuned using intelligent techniques/algorithms to search the parameter space quickly and efficiently. Multi-fold cross-validation techniques or inner-outer cross-validation techniques are often used to validate parameter selections. Optimization means finding the maximum or minimum of a function subject to specific constraints, so that the optimum parameters chosen result in maximum performance. It is well known that the parameters are subject-dependent, and should be selected for each subject. To optimize the set of parameters, various techniques are available including computational intelligence-based techniques such as particle swarm optimization (PSO) [252–254] or genetic algorithm (GA) [255]. A brief discussion of these techniques follows.

2.9.1 Cross-validation

Cross-validation is both an empirical and a heuristic approach typically carried out to assess how the results of a statistical analysis generalize over a set of independent data. There are several heuristics to choose the portions of the dataset to be used as a training and validation sets. In an m-fold cross-validation, the training set is randomly divided into m disjoint sets of equal size n/m, where n is the features in the data set. In BCI, cross-validation (CV) is carried out to obtain the most stable time point of the MI performed by the BCI user during the event-related time period across all the trials of a complete session. This is done to reduce the generalization error, because the classifier is chosen at the most stable time point and it is assumed to remain stable during successive sessions. Thus, the most stable time point is first obtained using the m-fold CV and thereafter the parameters of the classifier are chosen at this specific time point. In five-fold CV, the data are partitioned into a training set (80% of the data) and a test set (20% of the data). The procedure is performed five times using a different test partition (20%) data each time and a mean-CA and mean-Kappa coefficient is calculated from the CA and Kappa rates obtained from the five test partitions. Cross-validation is practical and a well-known method for evaluating the generalization performance of a classifier [185]. However, this approach necessitates heavy computation as the number of folds of the CV increases. Nevertheless, since it is an offline approach, this drawback does not matter much as long as the classifier generalization is achieved. Many BCI researchers commonly perform generalization through five-fold, 10-fold and inner-outer CV techniques, as discussed in references [17,44,180,220,256,257].

2.9.2 Genetic Algorithm

GA is an adaptive heuristic search technique derived from the idea of naturalization of the evolution process, i.e., survival of the fittest. It was pioneered by John Holland [255]. The pseudocode to implement GA is shown in Table 2.2. GA is started with a set of solutions known as chromosomes or genes or the population. At the end of one round of searching, a few of the solutions from the population of genes are taken and used to form a new population of genes according to their fitness. This means that the genes that are more suitable to produce better solutions are used for reproduction (through crossover and mutation).

TABLE 2.2   Pseudocode to Implement GA

Step	Explanation/Indication
1	Generate initial population of likely solutions (preferably within a specific search domain or region within which the member is likely to find an optimal solution).
2	Evaluate the fitness value (usually value of error) for all the members. In case of BCI, this parameter means improving the CA.
3	Choose the fittest members (or prune the worst) which can take part to form the next new generation of population.
4	Apply genetic operators (crossover, mutation) to form the new generation of population.
5	Check termination criteria (number of generations, minimum fitness threshold satisfied etc).
6	Go to Step 2 and continue the process if the termination criterion is not fulfilled.

This move from one generation to the next is motivated by an expectation that the new population will be better than the previous one.

The GA technique outperforms other traditional methods in most problems, as it is a more general and a flexible method with the capability of searching over a wide solution space and at the same time avoiding falling into a local minimum. Thus, it has the potential for finding an optimal or a near-optimal combination of the parameter values. This algorithm can perform a random search within a predefined search space and can solve a specific problem as well.

GAs differ substantially from traditional methods in many ways. GAs search a population of points in parallel and not as a single point; they only require the objective function that needs to be minimized or maximized and the corresponding fitness levels which influence the directions of the search. Also, GAs do not work on the parameter set directly, but on an encoded set (known as a gene) by using the search rules which are probabilistic in nature rather than the conventional deterministic rules. It is also important to mention here that the GA provides many solutions, but selecting a specific solution as the choice is at the discretion of the user. While the GAs can rapidly locate good solutions even for complicated search spaces, the technique has some disadvantages associated with it. The primary disadvantage is that the fitness function of the associated problem set should be defined properly, otherwise the GA may converge towards a local optima rather than the global optimum solution.

Figure 2.14 shows a graphical representation of the convergence of a population of GA over a number of generations to reach an optimum solution for the objective function: $f(x,y) = 6 * (x-y*y)^2 + (x-1)^2$. The final fitness value or error for a population of size 20 over 30 generations is 1.1201e$-$004 in reaching the best value of the function for variables x and y.

To conclude, it can be said that a major advantage of GAs is that they can quickly scan a vast solution set due to their inherent parallelism. Since the GAs work on their own internal rules, the genes giving a better solution are retained while the remaining ones are discarded in subsequent generations of evolution. This is a major advantage of GAs — they find optimal solutions through the process of evolution. However, in nature, life does not always evolve towards a better solution, and hence GAs may get trapped in a local minima, thereby giving a sub-optimal solution. A major advantage of GAs is that no prior knowledge about the problem is necessary. GAs can produce random changes (through crossover and mutation) in their solution genes and then use the objective function as a way to determine whether the applied changes are positive or negative to the overall search problem. The fitness function is crucial for an appropriate solution to the problem in GA. If the fitness

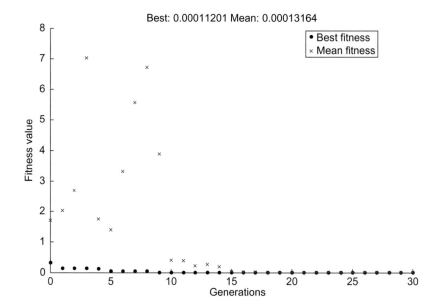

FIGURE 2.14   GA example.

function is inappropriately defined, then the GA may end up solving something different from the problem for which it was originally intended. It is also important to set the parameters such as crossover rate, mutation rate, population size, etc., which if defined improperly can result in the system never converging to a suitable solution. Thus, GAs should be applied to the specific problem by clearly identifying the parameters for the well-defined fitness function.

2.9.3 Particle Swarm Optimization

PSO is a heuristic-based stochastic optimization technique developed by Eberhart and Kennedy, and inspired by the social behavior of bird flocking and fish schooling [252–254]. PSO can easily deal with real world problems that invariably have some unknown parameters. PSO is similar to GA in many ways. Both the PSO and the GA methods are initialized with a population of random solutions. These random solutions are known as particles in PSO but as genes in GA. These particles search for optima in the search space and are updated as they attempt to reach the optimal solution. The GA and the PSO differ here in the update procedure leading to the new generation of genes or particles or commonly known as the new population. The GA has operators such as crossover and mutation parameters that help it to undergo the evolution

TABLE 2.3   Pseudocode to Implement PSO Technique

Step	Explanation/Indication
1	Initialize all the particles of the PSO with suitable values (preferably within a specific search domain or region within which the particle is likely to find an optimal solution)
2	Calculate the fitness value (usually value of error) for the particle. This may be interpreted as the classification accuracy for a BCI application
3	If the fitness value of this particle is better than the historical best fitness values lBest (local best), then replace the lBest with the new best fitness value
4	Choose the particle with the best fitness value amongst all particles as the gBest (global best)
5	Calculate the particle velocity for each particle and update the particle's position
6	Go to Step 2 and continue the process if the maximum no. of iterations is not reached or the minimum error criterion is not fulfilled

process. Contrary to this, the PSO has no such operators. Rather the PSO reaches towards potential solutions by stochastically adjusting the velocities of the individual particles according to the historical best position for the particle itself and the neighborhood best position. Here both the particle and the neighborhood best positions are derived according to a user-defined fitness function. PSO is not affected by the size and nonlinearity of the problem, and can converge to the optimal solution in many problems where most analytical methods fail. The advantages of PSO over GA are that the PSO is easy to implement and there are fewer parameters to adjust. Table 2.3 details the pseudocode to implement the PSO technique.

The information-sharing mechanism in PSO is thus considerably different from that in GA. In GAs, the chromosomes share information with each other so that the whole population moves towards an optimal region of solution. In the PSO, only the gBest (or lBest) are responsible for giving the information about the optimal region to the other particles within the group. Thus, there is a kind of one-to-many type of information-sharing mechanism in the PSO. The pseudocode described above for PSO has been applied in several optimization problems. A better initialization in the form of a probable best search region to find the optimal solution as well as particle velocity update and fitness evaluation can be specified to achieve an optimum solution more quickly.

Figure 2.15 shows a graphical representation of the convergence of particles over a number of generations to reach an optimum solution for the objective function: $f(x,y) = 6 * (x - y * y)^2 + (x - 1)^2$.

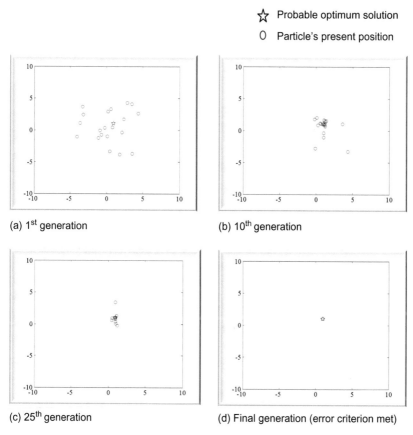

FIGURE 2.15 PSO example.

There are many PSO variants, such as constant inertia weight [254], time-varying inertia weight [258], constriction factor [259], etc., which may also be employed. PSO and GA are used for obtaining the optimal parameter set of the RQNN model.

2.10 GRAPHICAL USER INTERFACE [GUI]

The output signal from a BCI is limited and cannot be directly interfaced with technologies that are controlled using conventional means. With a simple two-class BCI system, there are only two output classes in the form of a left hand MI or a right hand/foot MI for every trial. With this binary signal and the limited capacity of BCI, the task of controlling various applications requiring multiple commands is a big challenge. Many of these BCI applications include word processors, games, wheelchairs/robotic control, etc. A graphical user interface (GUI) is one option that can enhance the information transfer to enable the use of the limited bandwidth of the BCI for controlling devices.

2.10.1 The Necessity for an Interface

The challenging issue of the limited bandwidth of a two-class BCI system can be considered by thinking of it as synonymous with selecting/controlling multiple controls for a robotic movement by using only two keys. A possible option for solving such a tricky situation is to use a multiple class BCI, for example, a three-, four- or eight-class BCI. However, the classification accuracy progressively reduces as the number of classes increase. If the control task requires four-class information (e.g., robot, each MI mental task of left hand, right hand, feet and tongue for a forward, left, right or backward control [185,260,261]), then the probability of error is 0.75, whereas if a two-class BCI is used for the same task then the error probability is lower, at 0.50. This increase in the error probability with increasing number of classes results from the increasing overlap of features in the feature space. Every classification of a mental task is afflicted with a probability of misclassification due to the overlapping features in the feature space. Increasing the number of mental tasks therefore increases the areas of overlap and increases the difficulty for the classifier [226], thereby decreasing the classification accuracy [260]. However, in contrast to the decrease in the classification accuracy, increasing the number of mental tasks simultaneously leads to an increase in the Information Transfer Rate (ITR)[13]. However, this might not justify the added complexity in terms of protocol design [45] as it requires more mental tasks. Improving classification accuracy should be the first target [45]. It is therefore proposed to use a two-class classifier to control a multiple task control problem and take the advantage of the higher accuracy rather than use a complex multi-class classifier with reduced accuracy. The BCI block diagram shown in Figure 2.16 focuses on the feedback and the interface portion to understand the dilemma graphically. The acquired EEG signal for a single trial can be termed as a left hand MI or a right hand MI (two-class). However, with this limited output information (i.e., the only information is that the user has performed one of the MI tasks), under many circumstances it is not possible to control a multi-tasked robot. It is

---

[13]ITR is the amount of information transmitted per unit time. It is derived from Shannon and Weaver [375] (summarized in Pierce [376]) and incorporates both speed and accuracy in a single entity.

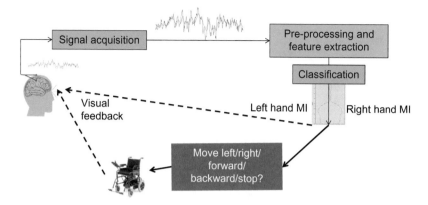

FIGURE 2.16   Need for an interface.

challenging for the robot to interpret the MI as a 'move left', 'move right', 'move forward', 'move back' or a 'stop/halt' command.

The issue can be understood by looking at the simple example shown in Figure 2.17. This shows a robotic arena, where the task is to maneuver the mobile robot from position A to position B with a simple two-class MI. Making the assumptions that 1) the mobile robot continually moves forward, 2) it incorporates a collision avoidance system and 3) it stops at every junction to receive a command from the BCI user, the user is expected to issue a decision/command to the mobile robot at the two junctions marked as J1 and J2 by associating one MI with one control task. At junction J1, the user can maneuver the mobile robot in a left or a forward direction through the commands interpreted from a simple two-class MI. Assume that the user decides on a forward movement by performing a right hand MI at J1 as indicated by the robot trail (cf. Figure 2.17(b)) and reaches junction J2 (cf. Figure 2.17(c)). However, the issue becomes complicated at junction J2, where the robot has three possible directions/outcomes (left, right and forward) and the user can only issue commands by using a two-class MI. Under such circumstances, the presence of an interface becomes a necessity in order to interpret and associate a two-class MI in the form of any user-intended command.

Consider another case in the same arena, where, because of some error/mistake, the robot has traversed to the left from junction J2 (cf. Figure 2.17(d)). It is now impossible to correct this error/mistake and maneuver the mobile robot back onto the desired track using the simple two-class MI-based interface. This is because the usual interfaces for robot control will now only stop at the next junction (wherever this is) for the user input. A good interface (as the one presented in this book) can handle such issues and suitably associate an MI with any

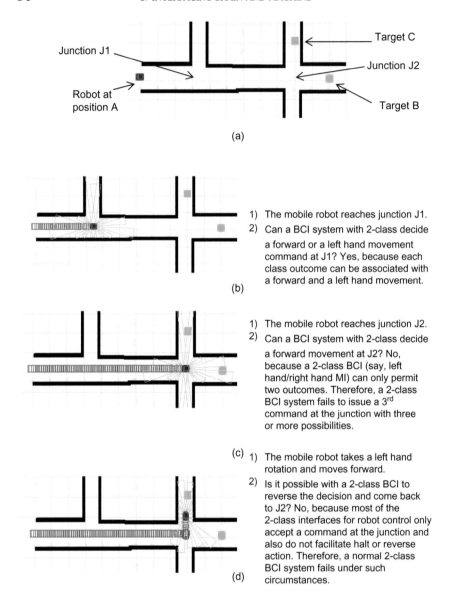

FIGURE 2.17   Challenges in maneuvering a mobile robot with a two-class BCI system.

desired possible outcome during the course of the movement of the mobile robot.

The purpose of this discussion was simply to justify the necessity for a good interface design that can potentially maneuver a mobile robot by using a two-class MI and simultaneously benefiting from its

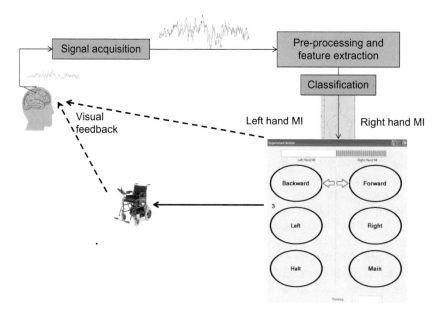

FIGURE 2.18   Interface solves the problem.

inherent higher accuracy. The answer to this kind of a dilemma is a suitable interface design. Revisit the block diagram shown in Figure 2.16 with a suitable interface placed in the form of visual information for the BCI user as shown in Figure 2.18. A horizontal arrow points to the two options marked 'backward' and 'forward'. This horizontal arrow continually translates to the next two options after every trial, typically of 8 s duration. In the present state, if the subject performs a right hand MI, then it is interpreted that the subject wants to select the 'forward' command. If the user had performed a left hand MI, then this would have been interpreted as a 'backward' command for the controlled device. If the subject does not perform any MI then the horizontal arrow, as per the trial timing, moves to the next two options. Therefore, the same left hand and right hand MI are now interpreted as a 'left' and a 'right' command if the user performs the MI when the horizontal arrow has the two options as 'left' and 'right;' i.e., if the horizontal arrow points between a different set of options. Therefore, this kind of approach has the potential to give another dimension of choice to the user, by associating the same two imageries (i.e., limited bandwidth) with different tasks by using a translating arrow within an interface. Thus, a good interface can serve the purpose of maneuvering the controlled device in a more efficient way by utilizing the limited bandwidth of a two-class BCI. The design of this interface is discussed in depth in Chapter 4.

The next sub-section discusses the features to be expected from a good user interface design.

2.10.2 Expectations from a Good GUI Design

Some important issues should be accepted and dealt with appropriately while designing an interface for the BCI user. A list of the features that one should expect from a good interface is given below.

a. The interface should be clear and concise. The information and choices should be conveyed quickly and clearly to the user (without any ambiguity).
b. The layout and interface design should be consistent.
c. The interface should adapt itself to compensate for BCI signal processing errors, i.e., despite the known problems of lower BCI accuracy, the interface should be such that it compensates for these issues. (This is discussed in depth in Section 4.2.1.2.)
d. The interface should be such that even the most likely events as well as the most unlikely events should be available as choices to the BCI user. If this is not the case, i.e., if some unlikely events are not offered as the available choices to the user, then the fundamental requirement of providing independence to the BCI user is lost, as this now solely depends upon the device/machine being controlled.
e. Good human−computer interface (HCI) design principles and accessibility should be considered.

Next is a list of issues that must be considered from the perspective of the BCI user.

a. The user may have very limited skills and may not be experienced.
b. Ideally, the user gives minimum effort and concentration time. This may be because the user may become tired through the continuous issuance of commands (and many of these may not be interpreted correctly due to the inherent drawback of the BCI signal processing) and thereby lose concentration [262−264]. Thus, the user will want to give minimum directions to the controlled device but at the same time should not lose control over the device by way of complete automation. This is a tricky issue and should be handled with extreme care.

To summarize, the effort (or cost) from the BCI user should be minimized by way of automation or semi-automation of the device to be controlled. Second, care should be taken to efficiently and automatically deal with incorrect decisions, i.e., when the user's intent is incorrectly interpreted by the BCI.

With these issues in mind, the next section describes recent developments in the GUI designs by some of the prominent BCI groups.

2.10.3 Recent Developments in BCI GUI Design

The design of the GUI depends on whether it is intended to work in tandem with a P300- (cf. Section 2.4.2) or an MI- (cf. Section 2.2) based BCI system. In addition, the design may also vary depending on whether it is scheduled for a synchronous BCI or an asynchronous BCI system. Several paradigms/interface designs have been proposed in the literature with the goal of increasing the communication bandwidth between user and device for effective control. The P300-based GUI designs require the user to select a particular choice in a more inherent way by only attending to the desired choice that is displayed amongst the different options, whereas the MI-based approach requires the user to actually select a desired choice by intentionally modulating their sensorimotor rhythms (SMR). Hence, the designs based on these two approaches differ extensively. A number of recent developments using P300- [5] and MI- [74,265,266] based EEG signals are discussed below. The focus of this section is to understand these GUIs and the major challenges faced in their practical implementation for speller and robot control tasks. This section is divided into interface designs for speller and robot control applications.

2.10.3.1 *GUI Designs for Speller Applications*

Some of the popular GUI designs for speller applications are discussed below.

2.10.3.1.1 P300-BASED VIRTUAL KEYBOARD

This P300-based interface design, similar to the one shown in Figure 2.19, has been developed using key selection [100,267,268] for a spelling application. The design has a menu with different keys (as options) displayed in front of the user. The P300 signal is the measure of user attention, rather than a direct visual signal [269], hence the interfaces based on these signals require the user to focus or count the number of times the target flashes [19,270], in order to detect the particular P300 that is characteristic of the desired task. These types of P300-based designs have remained almost the same over the past decade. The major issue with these designs is that the keys/objects need to be flashed at a specified interval in order to provide a means of generating the VEP or the P300 response from the BCI user. This results in known problems of eye gaze fatigue due to continuous visual focus.

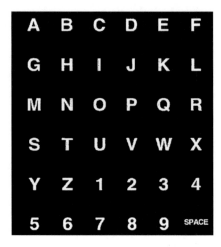

FIGURE 2.19   P300-based virtual keyboard. *Reproduced with permission from [100].*

FIGURE 2.20   Virtual keyboard during the writing of a message. The keyboard is divided into three segments (see top left panel), each associated with a different mental task and using the same colors as during the training sequence. The neural classifier's recognition of the same mental task three times in a row selects the corresponding segment of the keyboard (see top center); the green area is shadowed for 3.5 seconds to allow the user to undo the selection. This segment is divided again (see top right). A selected block is split in three again to offer a choice amongst the three letters (see bottom left). After the user selects the letter shown in red (h), the whole process starts over again (see bottom center). *Reproduced with permission from [74].*

2.10.3.1.2 MI-BASED VIRTUAL KEYBOARD

This MI-based keyboard interface design is described in reference [74], and a representation is shown in Figure 2.20. The MI approach requires the user to essentially select a desired choice in an intuitive way. In this design, as the users decide what they want to write, the

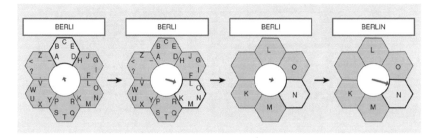

FIGURE 2.21   The mental typewriter 'Hex-o-Spell'. The two states classified by the BBCI system control the turning and growing of the gray arrow respectively (see also text). Letters can thus be chosen in a two-step procedure. *Reproduced with permission from [275].*

keyboard is split into three parts, each associated with one of the mental tasks the classifier has been trained to classify. Then, as the interface recognizes which task the subject is concentrating on, the keyboard splits successively into smaller segments until a single letter is finally selected, and then the process is repeated. This approach is effective but requires 22 s to select a particular letter, which is too slow.

2.10.3.1.3 MI-BASED HEX-O-SPELL TYPEWRITER INTERFACE

An efficient mental hex-o-spell typewriter interface with two mental states that uses a probabilistic text entry system has been reported in references [265,266,271–275] and is displayed in Figure 2.21. Here, six hexagonal fields surround a circle and six different selections/choices are arranged within each of these hexagons. There is an arrow in the center of the circle for the selection of a letter/symbol.

Imagining a right hand movement turns the arrow clockwise, and imagining foot movement stops the rotation and extends the arrow. If this imagining is performed for a longer period, the arrow touches the hexagon and thereby selects it. A language model [276] has been used for positioning the letters in the inner hexagons. This approach provides substantial increase in the ITR, but there are anecdotal reports of patients who preferred a slower spelling system to using a system that suggested word completions based on a probabilistic model [266].

2.10.3.2 GUI Designs for Robot Control

This section discusses the interface designs for mobility control applications.

2.10.3.2.1 P300-BASED GUI WITH PREDEFINED FIXED LOCATIONS

This P300-based GUI for controlling the movement of a wheelchair using a predefined set of locations was implemented in the studies

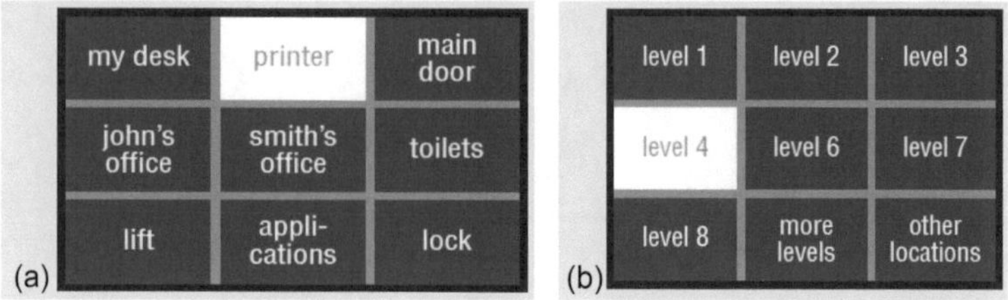

FIGURE 2.22   Context-dependent menus of commands displayed to the subject. The items flash in a random sequence onscreen, and the P300 BCI selects the item on which the subject is focusing. (a) This menu corresponds to the wheelchair navigating on one floor. (b) This menu corresponds to the location of the wheelchair at the elevator's 5th floor, which is thus not displayed. Note that the number of commands isn't limited to nine. *Reproduced with permission from [278].*

of references [5,269,277,278] and a representation is displayed in Figure 2.22. Here, a context-dependent menu is displayed in front of the user, various pre-designated items/locations are flashed in a random sequence onscreen, and the user is required to attend or focus to the item to be selected. Thus, if the user intends to select the target 'lift', then the user should focus attention on the icon 'lift', preferably by counting the number of times that icon flashes. With this kind of an interface and wheelchair control application, only predefined locations are available. Therefore, such an interface will only work within a known environment and is thus very limited in terms of providing the true independence needed by the user when the user wishes to move/control in any environment.

2.10.3.2.2 MI-BASED ROBOT CONTROL INTERFACE

A self-paced MI-based GUI has been investigated by Geng et al. [80]. In this study, as shown in Figure 2.23, there are three different-colored cursors representing three different tasks and these rotate continuously inside a circle. The user has to imagine foot movement to select the desired task when the corresponding cursor is within a designated command area. Since the movement of all the three cursors is visible at all times, the user is able to predict the time when a particular cursor will traverse within the command area. Due to MI latency, the user must start to perform the foot MI before the cursor actually reaches the command area. The advantage is that each of the three options is available to the user every 2−4 s and the user does not have to wait for a trial. However, in addition to the normal MI skills, the user must learn to modulate their SMR so that the threshold is reached at the time the target cursor has rotated to the command area. This is challenging,

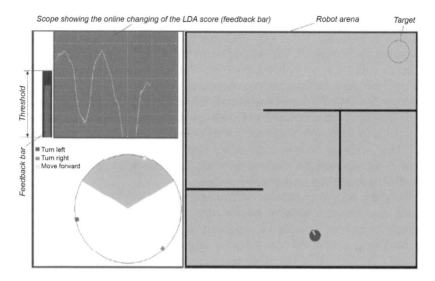

FIGURE 2.23   Interface paradigm with motor prediction for robot control experiment. *Reproduced with permission from [80].*

therefore, there is a need for an interface design that is simple and at the same time enhances the limited communication bandwidth. The major limitation of all the above-mentioned approaches is that the MI-based interface designs cannot give a control within less than 2 s due to the inherent latency of the mental imagination approach.

2.11 STRATEGIES IN BCI APPLICATIONS

A BCI can be used to perform tasks such as robot movement control, manipulator control, wheelchair control or to play games by issuing commands by thought through the GUI. However, GUIs need to be specific to a particular application or able to adapt to present the user with the available options. The following section discusses the strategies in BCI applications.

2.11.1 Shared Control BCI System

The well-known concept of shared control in BCI [279,280] enables the cooperation between a human and an intelligent device that allows the subject to focus the attention on his final destination and ignore low-level problems related to the navigation task (i.e., collision and obstacle avoidance) [281]. This approach has been successfully implemented in

[281,282]. Shared control methodology is being investigated through two general approaches within the BCI community: autonomous and semi-autonomous modes. In the autonomous approach, the subject interacts with the robot just by indicating a command representing the final destination and the robot decides the best trajectory to reach it. Thus, in this mode, full control of autonomously maneuvering the robot or wheelchair lies with the robot and not with the BCI user. A semi-autonomous approach is therefore preferred in BCI, to keep more control with the BCI user. In this semi-autonomous approach, full control of autonomously maneuvering the robot or wheelchair lies with the BCI user and the shared strategies only act as a guide.

Recently, adaptive shared control systems have been investigated in which the system utilizes information from the user, and simultaneously accesses information from the various sensors attached to the device being controlled, in order to improve the reliability of the BCI system. Thus, these systems are one step above the shared control strategies. These systems include algorithms such as obstacle detection, collision avoidance and orientation recovery for a brain-actuated wheelchair [282,283]. Collision avoidance is meant to decrease the translational velocity of the wheelchair until it comes to a halt if the user steers too close to an obstacle. On the other hand, obstacle avoidance adjusts the translational and rotational velocity until the controlled device/wheelchair is steered away from the obstacle [283]. Thus if the end task is known to the wheelchair in terms of a map layout, then the use of an obstacle avoidance approach will lead the wheelchair towards the target location, whereas the collision avoidance approach will halt the wheelchair at the location where there is a probability of collision. The orientation recovery algorithm corrects the orientation of the wheelchair if it gets misaligned with reference to the desired direction [282]. It is worth mentioning that, with the introduction of the orientation recovery, the behavior of the controlled wheelchair also changes. This is because the orientation algorithm reorients the wheelchair whenever it is off course and hence the subjects need to get accustomed to the altered behavior of the wheelchair.

The shared control approaches discussed above are used for maneuvering the robotic device, and therefore should not be considered as a methodology within the actual BCI framework. The usual shared control approach is utilized in this book (cf. Section 4.2.1) where the BCI user issues commands through the brain–robot interface (commonly referred to as the GUI) and the controlled device steers in response to the command, but within the framework of the inbuilt collision and obstacle avoidance. However, in addition to this, the strategy adopted within this book is a step ahead of the usual shared control methodology and is referred to as the intelligent Adaptive User Interface (iAUI) within the framework of the adaptive shared control BCI system (cf. Chapter 4). The iAUI presented in this book utilizes the information from the environment surrounding the device to be controlled and updates the iAUI in real-time (cf. Section 4.2.1), thereby enabling the BCI user to issue the required information in the form of a more likely command to be sent to the device more quickly. Thus, an adaptive shared control mechanism is presented in this book.

2.12 PERFORMANCE MEASURES OF A BCI SYSTEM

This section discusses the complex and critical issue of measuring the performance of a BCI system. The performance of different BCI systems is difficult to measure and compare. However, there are many publications which have analyzed and compared the selection of performance quantifiers that may be considered when designing a practical BCI system [5,22,44,80,101,133,269,272,278,282,284,285]. These include CA, response time, ITR, amount of mental attention required and ease of use. The easiest and the simplest measure of performance of BCI systems is to calculate the error rate/accuracy. However, this measure of comparison is solely related to the signal processing aspect of a BCI system. Therefore, the performance of a communication system is usually assessed based on ITR, mutual information or bit rate (B) [286] as suggested by Wolpaw et al. [19].

The ITR [287] combines the speed and accuracy into a single entity. Hence most synchronous BCI systems focus on improving the ITR [287] as a measure of performance evaluation. The bit rate or bits/trial is defined as follows:

$$B = \log_2 N + P\log_2 P + (1 - P)\log_2 \frac{1 - P}{N - 1} \qquad (2.17)$$

where P is the probability that a particular choice will be selected and N is the number of possible choices.

Then the ITR in bits/min is defined as:

$$ITR = \frac{60}{\text{trial duration}} * B \qquad (2.18)$$

Thus, the ITR depends upon the number of choices, the probability that the choice will be selected as well as the length of time taken to make a specific choice, i.e., the trial duration. A popular BCI feedback paradigm known as the basket paradigm [274,276] using a synchronous BCI protocol obtained an overall mean ITR for all the subjects as 4.9 bits/min with an average trial duration of 3.1 ± 0.7 s (a range of 2.1 to 4.0 s) [274]. However, there are still concerns over the use of ITR as a suitable means for evaluating the performance for a self-paced or asynchronous BCI system [288]. In addition, the comparison in the form of ITR may also prove to be a difficult issue when the choices within a user interface change. This may usually be the case when the number of alternatives/options that the users can select may vary in each scan cycle (as is the case in the work presented in this book) or the rate at which the selection occurs [51].

Another performance measure in the form of kappa coefficient[14] has also been proposed for unifying different classification problems. The kappa is used for comparison between BCI systems with different numbers of classes, where it is hard to use CA for comparison [289]. A CA of 50% in a two-class problem is the same as a CA of 25% in a four-class problem, which makes fair comparisons difficult for problems with varying numbers of classes when CA is used as the quantifier. In addition, performance measure with CA becomes more ambiguous when certain classes occur frequently during a specific process. However, the kappa coefficient represents the proportion of agreement obtained after removing the proportion of agreement that could be expected to occur by chance [290]. Therefore, some of the obvious results of CA obtained due to chance are removed in the performance measure through use of kappa. For an N class problem, if the N classes occur equally with a probability of $1/N$, the relationship between the kappa coefficient k and the CA can be described as:

$$k = \frac{CA - 1/N}{1 - 1/N} \qquad (2.19)$$

Therefore, kappa is a better measure of the performance of the system. Typically, the value of kappa lies between 0 and +1. However, as discussed before, the performance quantifier in kappa is also more indicative of the signal processing issues and not the interface-related quantification.

For applications such as wheelchair or robot control using MI, Fitts' law [246] can be used [291] to assess performance. Fitts' law is important for designers of the GUI because it provides guidelines for designing the interface elements and where they should be placed within the design, e.g., those tasks where the user must position a mouse/cursor on a particular button on the screen. According to Fitts' law [246,292],

---

[14]The kappa agreement index is a statistical measure indicating inter-rater reliability, i.e., agreement between two individuals and was developed by Cohen [377] for psychology analysis.

the movement time (MT) (time taken to complete the task) and Index of Difficulty (ID) (cf. equation 2.21) have the following linear relationship:

$$MT = a + b * ID \tag{2.20}$$

Here, a and b are constants. The symbol a represents the reaction time of the user and b represents the inherent speed of the device. The unit of b is time/bits or s/bits. In this form, the reciprocal of b is in bits/s and is called the index of performance (IP) or bandwidth.

Consider a simple task of moving a mouse cursor from one position to another position or target that is represented by a small bar. D is the distance between the present cursor location and the target bar, W is the width of the final destination or target bar. The difficulty of the task to have the cursor reach the desired location depends upon the width of the target W and the distance D. To mathematically express this difficulty, the Shannon formulation [293] of the ID can be used, which is expressed in 'bits' as follows:

$$ID = \log_2 \left(1 + \frac{D}{W} \right) \tag{2.21}$$

Decety et al. [294] investigated Fitts' approach to confirm if it holds true for MI-based designs. Fitts' approach was utilized to evaluate the EEG-based computer interface design in references [295,296] and was also used for analyzing the non-adaptive form of the proposed interface design in reference [56].

Wolpaw's approach is also quite popular among the BCI users as a quantifier to measure the performance of the BCI, although both approaches (those of Wolpaw and Fitts) do not take into account the training imparted to the user/subject for a particular task.

Another performance assessment method has also been described, which introduces the concept of utility with reference to a P300 speller design in reference [297]. This concept has been formalized as the expected average benefit for the user over time. This notion is mathematically expressed as the ratio of the average benefit to the user for performing any correct command in the expected interval of time.

$$U = \frac{b_L}{T_L} \tag{2.22}$$

where b_L is the average benefit carried by any correctly spelled letter over the expected time T_L. As expected, a higher value of the ratio U will indicate that the BCI user was able to reach the desired target in the shortest interval of time.

Some of the methods discussed above are a measure of the performance of a complete BCI system. However, some of these approaches measure the performance of the BCI by calculating the ITR or the utility of a complete interface that is non-adaptive, i.e., an interface that remains fixed over the complete duration of usage by the BCI user. A parameter such as the ITR depends upon the accuracy of the classifier and is hence somehow dependent on the signal processing rather than the interface design. The utility metric concept that depends upon the average benefit to the user is a good indicator for P300-based designs. However, the interface presented in this book is MI-based, and its main goal is to improve the communication bandwidth. The approaches discussed above do not give a true picture of the benefits that an adaptive interface such as the one discussed in this book can offer. The approach undertaken by many researchers for brain-controlled wheelchair (BCW) interface design performance analysis [298,299] is to utilize some form of total time for task completion (also referred to as mission time) as well as the amount of time the user has to spend issuing commands (also referred to as the concentration time). The mission time is the time to select a destination or the target on the user interface plus the total traveling time needed for maneuvering the mobile robot to reach the target as interpreted in reference [300]. The concentration time is the mission time minus the relaxation period (within each trial duration) or the period of mental imagination within a trial. This measure can be made independent of the signal processing issues in BCI. Hence, the approach followed in this book to quantify the real assistance that a user gets by way of the interface adaptation approach (cf. Section 4.2.1.2) is carried out by utilizing the state-of-the-art BCI quantifiers such as mission time, concentration time and the total cost as proposed in reference [300]. The adaptive, non-adaptive and autonomous interface designs for mobile robot control presented in this book (cf. Chapter 4) are compared with the contemporary interface designs in a practical scenario in Chapter 6.

2.13 CONCLUSION

The general structure of the complete BCI system, from the signal processing aspects to the GUI design for issuing the ultimate command to the controlled device is discussed in detail here, along with some novel strategies for practical implementation of the complete BCI system and some performance evaluation measures. These include the signal acquisition process, preprocessing, feature extraction, classification, post-processing and the GUI display and utilizing the feedback information in the form of adaptive strategies in BCI.

A very important block within the complete BCI system is the preprocessing technique, which, as discussed earlier, has the potential to reduce the computational load on the rest of the BCI components. Hence, the work presented in this book includes contributions in the area of signal preprocessing in the form of quantum mechanically motivated RQNN models. The next chapter discusses the basic concepts of quantum mechanics and the proposed models (based on these quantum mechanical concepts) which are investigated in simple DC and sinusoidal signals for the filtering process.

Likewise, a good post-processing technique is also necessary as it can enhance the stability and the way in which the classifier output can be utilized for operating the controlled device. Similarly, a good interface design can enhance and speed up the controlling of the device. Hence, another major contribution in this book is in the form of a different way of using the classifier output through the post-processing block, while utilizing a dynamic and intelligent brain−robot interface. As discussed in Chapter 4, the proposed interface is designed to acquire the information from the surrounding environment in terms of the existing obstacles in all the directions of the device being controlled. Using sonar sensor information, the brain−robot interface reorders the display of choices in a chronological order (and highest priority is given to the choices with minimal obstacles in the corresponding path direction), thereby allocating the first choice to the user of the task that can be honored most conveniently. Collision avoidance and obstacle avoidance techniques are also utilized. These aspects have improved the existing shortcomings discussed in Section 2.10.1, and make the proposed interface paradigm user friendly, quicker and practically usable in unknown environments.

3

Fundamentals of Recurrent Quantum Neural Networks

3.1 INTRODUCTION

This chapter describes the architecture of the quantum neural network (QNN) as existing in the literature. In this framework, a layer of neurons which are spatially situated encode the stochastic features associated with a noisy signal to the potential function associated with the Schrodinger Wave Equation (SWE). Using an unsupervised learning scheme that applies to the weights of this spatial neural network, the potential function is modulated in such a way that the quantum state ψ, the solution of the SWE, carries the information regarding the time varying probability density function (pdf) associated with this input signal. The approach is particularly suitable when the nature of the input signal, as well as the nature of additive noise within the signal, is not known. In this sense, this QNN method provides an intelligent signal processing framework for stochastic filtering. Some simulation experiments have been carried out to validate this claim. Limitations of the existing QNN architecture have been enumerated as well.

3.2 POSTULATES OF QUANTUM MECHANICS

Quantum mechanics (QM) provides a mathematical description of quantum objects that exhibit dual particle and wave nature. The abstract mathematical formulation of QM is best understood through postulates. At the heart of this theory is the wave function that is defined in Hilbert space. This wave function provides probability amplitudes of

65

observables such as position and momentum. QM postulates are given as follows:

Postulate 1: The complete information about the particle is given by a wave function $\psi(x)$. In other words, the state of a QM system is completely described by a wave function.

Postulate 2: Given this description of the system/particle in the form of ψ, the probability of finding the particle between position x and $x + dx$ is given by $P(x)dx = |\psi(x)|^2 dx$.

Postulate 3: The behavior of $\psi(x, t)$ is governed by the SWE:

$$-\frac{\hbar^2}{2m}\nabla^2\psi(x, t) + V(x, t)\psi(x, t) = i\hbar\frac{\partial\psi(x, t)}{\partial t} \tag{3.1}$$

Each term in this equation has a specific meaning that will be discussed in the next section.

Postulate 4: The measurement of physical quantities (say, e.g., energy and momentum) depends on the behavior of the operators associated with these quantities. For every dynamic variable in classical mechanics, there corresponds in quantum mechanics a mathematical operator in conjunction with the wave function.

Postulate 5: For any dynamic variable, there exists an expectation (average) value that can be calculated from the wave function and the corresponding operator of the dynamic variable.

Detailed information about these postulates is available in reference [301]. Many statements are also found in literature which are often presented as postulates, however, these are only the consequences of the above five postulates. The next section discusses quantum mechanics and the SWE in more detail.

3.3 QUANTUM MECHANICS AND THE SCHRODINGER WAVE EQUATION

The concept of superposition is fundamental in QM. Assume a quantum state represented by $|\psi\rangle$ (which is a vector in the Hilbert space $\mathcal{H}$) and also referred to as a wave function. This quantum state $|\psi\rangle$ can be constructed from a superposition of base states. The wave function $|\psi\rangle$ is thus a mathematical depiction of a quantum system, and in a way it informs us of all the possible states in which the system can be located. The location of the system/particle is dictated by a wave equation that describes the probability of the particle's existence at any given point of time. Thus, in essence it is a collection of all probabilities. Although, whenever measured the system/particle has a definite location. A system can thus be constructed using many base states. For simplicity, consider a system with only two base states $|\psi_1\rangle$ and $|\psi_2\rangle$. Thus, the wave function $|\psi\rangle$ for such a system can be written as:

$$|\psi\rangle = \Theta_1|\psi_1\rangle + \Theta_2|\psi_2\rangle \tag{3.2}$$

Here Θ_1 and Θ_2 are arbitrary complex numbers. ψ_1 and ψ_2 are called the base states, also known as a qubit, i.e., an unobserved base state. Thus a qubit can have states $|0\rangle$, $|1\rangle$ and a linear combination of the states $\Theta_1|0\rangle + \Theta_2|1\rangle$ where $\Theta_1^2 + \Theta_2^2 = 1$. When the base state is observed, it is known as a bit, either 0 or 1. In this form, i.e., when the state is observed, the wave function is said to 'collapse' into a definite state, which is in fact the reality that one observes. The variables $\begin{bmatrix} \Theta_1 \\ \Theta_2 \end{bmatrix}$ are normally written in a matrix form and are known as the coefficients. It is interesting to note here that the information is contained within the coefficients and not in the base states. The base states are simply all the possible states in which the system can be located.

The time evolution of the state vector $|\psi_1\rangle$ is according to the SWE [302] and is represented as:

$$i\hbar\frac{\partial\psi(x,t)}{\partial t} = H\psi(x,t) \tag{3.3}$$

where H is the Hamiltonian operator or the energy operator and is given as $i\hbar\frac{\partial}{\partial t}$. The Hamiltonian is thus not a number or a function but is an operator, similar to any spatial operator represented as $\frac{\partial}{\partial x}$. $\psi(x,t)$ is the wave (probability amplitude) function associated with the quantum object at space-time point (x,t).

The above equation (3.3) can also be represented as:

$$E\psi(x,t) = H\psi(x,t) \tag{3.4}$$

where E is again the energy operator. The energy operator can also be understood from a different perspective.

The Hamiltonian essentially represents the total energy in the system [303] and is thus the sum of the kinetic energy T and the potential energy V,

$$H = T + V \tag{3.5}$$

However, $T = \frac{1}{2}mq^2 = \frac{p^2}{2m}$ where p is the momentum, q is the velocity and m is the mass of the object. In making the transition to a wave equation, i.e., to convert to QM, Schrodinger constructed the momentum operator *as* $\hat{p} \rightarrow -i\hbar\nabla$ from the physical variable, i.e., $\hat{p} = -i\hbar\frac{\delta}{\delta x}$. Here $\hbar$ is the reduced Planck's constant $(h/2\Pi)$ and h is Planck's constant (a constant denoting the size of quanta in QM). Thus, V takes the form of a function $V(x, t)$ of position and time. Therefore,

$$H = -\frac{\hbar^2}{2m}\nabla^2 + V(x, t) \tag{3.6}$$

where ∇^2 is the Laplacian operator[1].

It is thus possible to apply the Hamiltonian to systems described by the wave function $\psi(x, t)$, i.e., the Hamiltonian acts on the wave function to generate the evolution of the wave function in time and space.

The time-independent SWE in 1-D is represented as:

$$-\frac{\hbar^2}{2m}\nabla^2\psi(x) + V(x)\psi(x) = E\psi(x) \tag{3.7}$$

The wave function ψ may not depend on time, but will always depend upon space. As a function of space, it maps the possible states of the system into the field of complex numbers and the SWE describes how ψ evolves over time.

The SWE is the basic equation governing the dynamics of a particle in QM. The time-independent SWE can thus be elaborated (to include time) as:

$$i\hbar\frac{\partial\psi(x, t)}{\partial t} = -\frac{\hbar^2}{2m}\nabla^2\psi(x, t) + V(x, t)\psi(x, t) \tag{3.8}$$

Normally one wishes to get $\psi(x, t)$ which represents the solution of the equation. Once the functional form of $V(x)$ and $\psi(x)$ are known at time t_0, techniques for solving the differential equation such as the classical explicit method, Hopscotch scheme, Crank-Nicolson scheme, etc. [304] can be applied to get $\psi(x, t)$. The time evolution of wave functions can thus be considered to be deterministic, because given a wave function at an initial time, one can make an explicit prediction of what the wave function will be at any later time. This capability to predict the future behavior of the wave from its present spatial configuration is a very important property of the time-dependent SWE. However, this presumes that knowledge of the potential $V(x, t)$ as a function of space and time is available. The random or probabilistic characteristics

---

[1]The Laplacian operator is the summation of the second derivatives of a function with respect to each of its independent variables, in other words, a differential operator. It is expressed as $\nabla^2 = \frac{\delta^2}{\delta x^2} + \frac{\delta^2}{\delta y^2}$ for a two-dimensional function in the form of $u(x, y)$. Similarly, for a function $u(x, y, z)$ in three dimensions, the Laplacian operator is expressed as $\nabla^2 = \frac{\delta^2}{\delta x^2} + \frac{\delta^2}{\delta y^2} + \frac{\delta^2}{\delta z^2}$, where x, y and z are coordinates in the Cartesian space.

of QM arise during the time of measurement only. The solution $\psi(x,t)$ is:

$$\psi(x,t) = e^{-\frac{i}{\hbar}H(t-t_0)}\psi(x,t_0) \tag{3.9}$$

and $\psi(x,t)$ should thus capture the statistical behavior of the signal.

Reverting to equation (3.2), where the quantum state $|\psi\rangle$ was defined with only two base states $|\psi_1\rangle$ and $|\psi_2\rangle$. To recover the base state, one needs to observe the system. However, by observing the system, one can recover only one state. As far as Newtonian mechanics is concerned, this base state or the estimated value holds true irrespective of the observer. However, in QM, the observer also plays an important role. In other words, QM depends on representation, i.e., the estimated value differs from one observer to another. An example of a classical and a quantum register is discussed below to illustrate this point.

3.3.1 A Classical vs. Quantum Register

Suppose there are two qubits, i.e., unobserved base states. If these were classical bits, there can be four possible states; 00, 01, 10 and 11. Corresponding to this, a two-qubit quantum system will have four computational basis states denoted as $|0\ \ 0\rangle, |0\ \ 1\rangle, |1\ \ 0\rangle$ and $|1\ \ 1\rangle$. Since a pair of qubits can exist in superposition of these four states, the state $|\psi\rangle$ describing the qubits is represented as:

$$|\psi\rangle = \Theta_1|0\ \ 0\rangle + \Theta_2|0\ \ 1\rangle + \Theta_3|1\ \ 0\rangle + \Theta_4|1\ \ 1\rangle \tag{3.10}$$

where Θ_x are the coefficients. The measurement x (=00, 01, 10 or 11) occurs with a probability of $|\Theta_x|^2$. The information is thus contained in $\Theta_1, \Theta_2, \Theta_3$ and Θ_4 and not in 00, 01, 10 or 11. Since $|\psi(t)\rangle$ is different from $|\psi(0)\rangle$, the $\Theta_i(t)$ are also different from $\Theta_i(0)$; i.e., as time passes, the expansion coefficients also change.

The probability of finding the particle described by the wave function ψ at any spatial location (x,y,z) at time t is proportional to the value of $|\psi|^2$ at that time. Hence, normalization of the quantum state is important by imposing the condition:

$$\sum_{i=1}^{N} |\Theta_N|^2 = 1 \tag{3.11}$$

In a practical sense, this is interpreted as stating that the probability of finding the particle within the specified region is 1; i.e., the particle should be found somewhere in the space, if it exists. Thus, the wave packet only gives the probability that the particle it represents will be found at a given position.

3.3.2 Quantum Neural Network

According to the DARPA Neural Network Study (1988, AFCEA International Press, p. 60) [305], a neural network (NN) is defined as follows:

> "...a neural network is a system composed of many simple processing elements operating in parallel whose function is determined by network structure, connection strengths, and the processing performed at computing elements or nodes."

Thus, an artificial neural network (ANN) is an information-processing paradigm that is inspired by the way in which biological nervous systems, such as the brain, process information. This network is composed of a large number of highly interconnected processing elements referred to as neurons that work in unison to solve specific problems [306]. The work presented in this book for enhancing the noisy electroencephalogram (EEG) signal utilizes a layer of neurons in the spatial dimension within the neural network framework. The incoming noisy input signal sample is treated as a probability density function (*pdf*) by the layer of neurons and it recurrently evolves under the influence of the SWE and appropriate learning rules. This approach has made possible the development of an efficient computational algorithm referred to as the recurrent quantum neural network algorithm (RQNN) which to some extent has solved the complex problem under consideration.

In Figure 3.1, it is shown that an external stimulus in the form of a single sample can be represented as a 1-D array of states in the quantum space. In neural network theory, this 1-D array can be considered as an array of neurons in a lattice. In other words, the single sample external stimulus of the input signal reaches each neuron in a lattice (a spatial structure of an array of neurons where each neuron is a simple computational unit) with a probability amplitude. In QM, the carrier of the

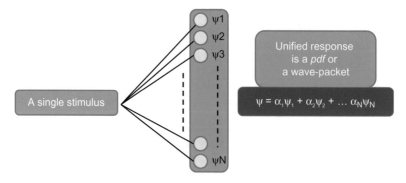

FIGURE 3.1   Theoretical model (quantum neurons).

stimulus performs quantum computation and hence the collective response of all the neurons is given by the superposition equation:

$$\psi = \alpha_1\psi_1 + \alpha_2\psi_2 + .. + \alpha_N\psi_N = \sum_{i=1}^{N}\alpha_i\psi_i \qquad (3.12)$$

and evolves according to the SWE. Assuming a negative probability [307] is not natural (since the interpretation of a negative probability is unknown), hence ψ cannot be an observable or a measurable quantity. In other words, as per some of the interpretations of QM, the wave function ψ is assumed to have no physical interpretation, as there is no reality attached to it. This creates a dilemma regarding the purpose of the wave function. This objection in the form of ψ not being an observable quantity cannot be applied to $|\psi|^2$ as it is the square of the absolute value of the wave function, i.e., the probability density or the probability of finding the particle at a given time. In addition, it also guarantees that only a positive quantity is referred to, because the square of any value is always positive. Hence, what is measured is the probability, and it can be related to ψ as $|\psi|^2 = \psi^*\psi$. Probability thus plays the role of a link between the object being a wave as well as a particle. It is worth mentioning here that given only $\psi^*\psi$, it is not possible to determine ψ.

Modern quantum theory should be viewed as a generalization of the well-established classical Newtonian mechanics. The time-dependent SWE is the basic dynamic equation of QM and generalizes the Newton's second law, Force = mass × acceleration. The SWE thus describes how the wave function changes over time. There is no way to derive the SWE from first principles; as truly speaking there are no "first principles" in physics that precede QM to predict, for example, the wave behavior for the electron. The SWE has thus to be postulated, just like Newton's laws of motion were originally postulated. The only justification for making such a postulate is that it works[2] [303]. With the time-dependent extension of the SWE, QM can now be expected to model situations that are not stationary [218].

3.3.2.1 Compelling Motivation Towards
the Quantum Filtering Approach

The brain processes responsible for thinking are fundamentally quantum rather than classical in nature [308]. This prompts the question of whether the human mind and mental processes can be considered

---

[2]As mentioned in [303], the reader should get used to this statement. Again and again, things are postulated in quantum mechanics, with the only justification being that it works.

inherently quantum in nature. In reference [309], Eliano Pessa argues that most mental processes should be described by quantum physics. The conceptual structure and the axiomatic foundations of quantum theory have a profound link with mental entities and their dynamics [310]. Intuitively, this indicates that designing alternative approaches in brain–computer interfaces (BCIs) with the amalgamation of quantum and classical concepts may lead to more efficient and robust bio-signal processing. QM is extremely successful in describing nature. However, the meaning of this description is still a matter of intense debate. The features believed to be relevant to the quantum mechanical model of the brain with reference to nondeterminism, contextuality and non-locality have been discussed in reference [311]. Busemeyer et al. [312] analyzed the dynamics of human decision-making and showed that a better fit of the data can be achieved using quantum dynamics. Quantum dynamics describe the evolution of complex valued probability amplitudes over time [312] and are derived from a slightly different set of basic rules, as discussed at the start of this chapter. According to quantum theory, the unobserved state transitions obey probability amplitude rules. As mentioned in the previous section, the general interpretation of QM thus associates the Schrodinger wave function in a probabilistic manner with $\left|\psi(x,t)\right|^{2}$ which acts as a probability density rather than in the classical deterministic way where there are explicit predictions about a specific outcome.

It is said that EEG signals are a realization of a random or stochastic process [32]. Thus when an accurate description of the system is not available, a stochastic filter can be designed based on probabilistic measures. Bucy, in reference [313], states that every solution to a stochastic filtering problem involves the computation of time varying *pdf* on the state space of the observed system. Using this concept, Dawes in references [314, 315] proposed a novel model – a parametric avalanche stochastic filter and in the progression confirmed Bucy's statement. The work on EEG signal enhancement detailed in this book is strongly inspired by these contributions, as it also involves the computation of a time varying *pdf* by heavily relying on the principles of QM and the SWE. Considering the neural lattice as representing a probability amplitude function mentioned earlier, the neural network theory and topologies can be used to implement a QM-based filtering approach, referred to as a RQNN. In the architecture of the RQNN model, each neuron (within the NN model) mediates a spatio-temporal field with a unified quantum activation function that aggregates the *pdf* information from an observed noisy input signal. The solution of the SWE gives the activation function. From a mathematical point of view, the time-dependent, single-dimension, nonlinear SWE is a partial differential equation describing the dynamics of the wave packet in the presence of a potential field (or function) (which is the force field in which the particles defined by the wave function are forced to move) [316]. Thus, the RQNN approach is based on QM and does not make any assumption about the nature and shape of the noise that is embedded within the signal to be filtered. In a way, the RQNN-based approach is considered more suitable for those systems where knowledge of the embedded noise characteristics within the signal are not known. EEG signals are one such type of signal where the characteristic of the embedded noise is not known. Thus, this network is utilized to model dynamic systems in order to fill up the void created by the non-availability of any *a priori* knowledge or information about the type of noise being dealt with when processing the biological signal. In summary, the RQNN is required to answer positively the following intriguing question.

Can a model estimate and remove the embedded noise in a non-stationary EEG without assuming any knowledge about the behavior or the amount of noise?

With the motivation for working towards an RQNN-based filtering technique established, the next section details the basic theoretical concepts that develop this theory.

3.4 THEORETICAL CONCEPT OF THE RQNN MODEL

John Searle declares that mental phenomena including thoughts, indeed all mental life, are caused by processes going on in the brain [317]. Therefore, intuitively, one can think of the brain as a hierarchical system whose mental part, modeled as a quantum process, observes the unified response of a specific neural lattice and actuates a feedback signal [318]. Based on this, the RQNN ignores the individual neuronal dynamics and provides a framework which may represent this quantum hypothesis of the unified behavior of the human brain in response to various types of stimuli [319]. The time evolution of this unified behavior ψ is described by the SWE [302], as mentioned previously.

The purpose of the neuronal lattice is basically to set up a spatial potential $V(x)$. A quantum process, described by the quantum state ψ which mediates the collective response of the neuronal lattice, evolves in this spatial potential $V(x)$ according to the SWE. As discussed earlier, $V(x)$ sets up the evolution path of the wave function and this is analogous to tuning the value of force being applied to attain a controlled movement of any physical object in the classical world based on Newton's law. Thus, any desired response can be obtained in the quantum model by properly modulating the potential energy. The quantum state is initialized to a state, say $\psi(x, t_0)$, with a mean value of zero and

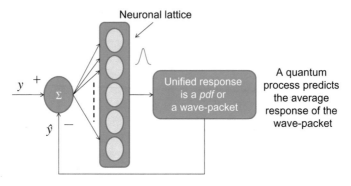

FIGURE 3.2   Conceptual framework of RQNN. *Reproduced with permission from [213].*

as time progresses from t_0 to say t_n, the quantum state also evolves in the spatial potential field $V(x)$. In other words, given the wave function at some time $t = 0$, $\psi(x,0)$, the SWE will give $\psi(x,t)$ at all times larger than $t = 0$.

It is known that recurrent networks are effective in dealing with time-dependent signals. This is the basis for designing the conceptual framework of the functioning of the RQNN model as given in Figure 3.2. It is similar to the one discussed in the previous section (cf. Figure 3.1). The model is basically a one-dimensional array of neurons whose receptive fields (i.e., neurons) are initially excited by the signal input y reaching each neuron through the synaptic connections. The array of neurons is arranged in the spatial dimension, thereby creating a lattice formulation. It has been shown in references [318, 320] that a quantum process models the average behavior (collective response) of a neural lattice. The lattice can be envisioned as if some quantum object in the form of every neuron in the neuronal lattice is expecting an incoming signal sample that is within its range. This specified range of the neuronal lattice can be set depending on the range within which the particle is expected to be located[3]. Each of these neurons within the lattice can be spaced at a specific resolution[4]. In addition to this, the number of neurons in the

---

[3]If the particle is expected to lie within a range of -3 to $+3$ then the neurons within the spatial layer should be set accordingly.

[4]E.g., if the values of two particles are say 0.31 and 0.33, then it is possible to locate them at different neurons only with a resolution of the neuronal lattice of 0.01 or lower, i.e., there should be a neuron each at the location 0.31 and 0.33. The said particles cannot be discriminated with a neuronal lattice resolution of 0.1 as both the values in the form of 0.31 and 0.33 will be found at the neuron located at 0.3.

lattice can also be fixed[5]. The lattice can thus be assumed to have a fixed number of neurons with a definite spacing between them to manage a suitable range of the input signal. The neurons in the spatial lattice respond to the input signal in the form of a wave packet. The recurrent evolution of this wave packet is according to the difference between the noisy input signal and the filtered estimate. If the statistical mean of the difference is zero, then the noisy signal will have little effect on the movement of the wave packet and precisely the movement of the wave packet along the desired direction will be due to the actual signal content within the noisy input signal. This is the goal of this technique.

The next section details the signal preprocessing/signal enhancement using the RQNN model.

3.5 TRADITIONAL RQNN-BASED SIGNAL ENHANCEMENT

In the RQNN, the hypothesis is that the average behavior of the neural lattice that estimates the true input signal is a time varying *pdf* (referred to later as $\rho(.)$), which is mediated by a quantum object placed in the potential field modulated by the input signal, thus transferring the information about the *pdf*. The SWE is used to track this *pdf* function.

Figure 3.3 shows raw (unfiltered/noisy) signal fed to the RQNN in the form of $y(t)$ and a filtered signal is obtained in the form of $\hat{y}(t)$ as the output of the RQNN model. The raw signal can be any signal with embedded noise that needs to be filtered out. Figure 3.4 shows the graphical understanding of the evolution of the wave packet at different

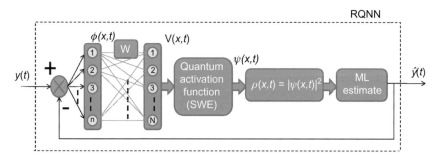

FIGURE 3.3   Signal estimation using the RQNN model.

[5]If the range of the neuronal lattice is -3 to $+3$, then with a spacing of 0.1 between each neuron, the total number of neurons each located at $-3, -2.9, -2.8 \ldots -0.1, 0, +0.1, \ldots 2.9, 3$ will be 61.

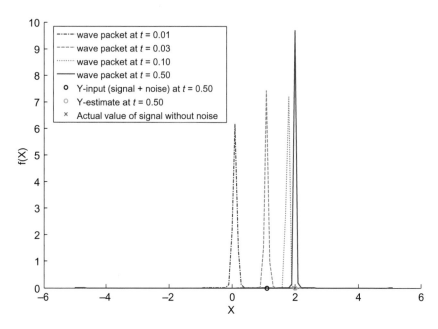

FIGURE 3.4   Evolution of the wave packet and estimate of the filtered signal.

times and the final estimated value of the filtered signal. The raw signal fed to the RQNN model is embedded with unknown noise. This noisy signal is treated as a wave packet that evolves (cf. Figure 3.4 in which the wave packet is shown at times $t = 0.01, t = 0.03, t = 0.10$ and $t = 0.50$) and finally stabilizes at a Maximum Likelihood (ML) estimated value of two, which is in fact the true estimate of the filtered signal.

The potential energy is calculated as below:

$$V(x, t) = \sum_{i=1}^{n} W_i(x, t)\phi_i(v(t)) \tag{3.13}$$

where $\phi_i(.)$ is a Gaussian kernel function and n represents the number of such Gaussian functions to describe the nonlinear map that represents the synaptic connections with time-varying synaptic weights $W_i(x, t)$. $v(t)$ represents the difference between the input signal y and the true estimate $\hat{y}$ and is represented as $v(t) = y(t) - \hat{y}(t)$.

The traditional RQNN model presented here has the neurons stimulated with the error signal [36, 37]. The Gaussian kernel function is:

$$\phi_i(v(t)) = exp(-(v(t) - g_i)^2) \tag{3.14}$$

where g_i is center of the ith Gaussian function ϕ_i. This is chosen from the input space by uniform random sampling. When a particle has its

wave function given by a Gaussian wave packet, the uncertainty prod-
uct has its minimum value. (With a wave packet whose shape is different from a Gaussian, an uncertainty principle applies that has the form
$\Delta pos\ \Delta p = K\hbar$ where pos is position, p is momentum and K is a numerical constant that depends on the precise shape of the packet. With a
Gaussian packet, one gets the smallest possible value of K, which is $1/2$.)
Hence, the Gaussian is preferred over other wave packets.

Since the parameters of the distribution are unknown, it is best to
estimate the same. The standard approach to this problem is the maximum likelihood estimation (MLE)[6] [321]. Therefore, the filtered estimate
of the signal is calculated using MLE as:

$$\hat{y}(t) = \int x(t)\rho(x,t)dx \tag{3.15}$$

Based on this estimate, the weights are updated (as discussed below)
thus establishing a new potential for the next time evolution.

The closed form dynamics of this recurrent quantum neural network
thus becomes:

$$i\hbar\frac{\partial\psi(x,t)}{\partial t} = -\frac{\hbar^2}{2m}\nabla^2\psi(x,t) + \zeta G(y(t) - \int x\psi(x,t)^2dx)\psi(x,t) \tag{3.16}$$

where $G(.)$ is a Gaussian kernel map introduced to nonlinearly modulate the spatial potential that excites the dynamics of the quantum
object, and ζ is the scaling factor that is utilized to actuate the spatial
potential $V(x)$. In fact, $\zeta G(.) = V(x,t)$, where $V(x,t)$ is calculated from
equation (3.13).

When the estimate $\hat{y}$ is the actual noiseless signal, then the signal that
generates the potential for the SWE $v(t)$ is simply the noise that is
embedded in the signal. If the statistical mean of the noise is zero, then
this error-correcting signal $v(t)$ has little effect on the movement of the
wave packet. Precisely, it is the actual signal content in the input $y(t)$
that moves the wave packet along the desired direction which in effect
achieves the goal of the EEG signal estimation. It is expected that the
synaptic weights evolve in such a manner so as to drive the ψ function
to carry the exact information of the pdf of the filtered signal $\hat{y}(t)$. To
achieve this goal the weights are updated using learning algorithms, as
discussed below.

The learning algorithms are based on changing the synaptic weights
between the neurons by specific learning rules. Three rules have been

---

[6]In statistics, MLE is a method of estimating the parameters of a statistical model.

TABLE 3.1   Pseudocode for Implementing the RQNN Model

Step	Explanation/Indication
1	Initialize the RQNN after declaring suitable parameter values (as discussed in Section 3.4)
2	Feed the next sample $y(t)$ to the RQNN model
3	Calculate the error $v(t)$ from the value of $y(t)$ and the estimate $\hat{y}(t)$
4	Calculate the value of the potential $V(x, t)$ from the updated value of $\phi(v(t))$
5	Calculate ψ and obtain the pdf $\rho(x, t)$ and then estimate (after normalizing the pdf) in terms of $\hat{y}(t)$ using MLE
6	Calculating the value of error and then update the synaptic weight vector $W(x, t)$
7	Repeat steps 3−6 for a fixed number of iterations (say MAXIT) to get the filtered estimate of the input sample
8	Go to Step 2

investigated within this RQNN model and these are discussed below. The first learning rule to update the synaptic weights is:

$$\frac{\partial W_i(x, t)}{\partial t} = \beta\phi_i(v(t))\rho(x, t) \tag{3.17}$$

where β is the learning rate.

The synaptic weights W can also be updated using the well-known Hebbian learning rule [322].

$$\frac{\partial W_i(x, t)}{\partial t} = \beta\rho(x, t)v(t) \tag{3.18}$$

3.5.1 Pseudocode for the RQNN Model

The RQNN modeling approach discussed in the previous subsection can be implemented using the pseudo code detailed in Table 3.1.

3.5.2 RQNN Parameters

There are many parameters to set in the RQNN model. Of these, three major parameters should be tuned appropriately in order to obtain the desired response of the SWE. The SWE and its nonlinear form exhibits a soliton[7] property for some range of parameters. There are

[7]In mathematics and physics, a soliton is a self-reinforcing solitary wave (a wave packet or pulse) that maintains its shape while it travels at constant speed.

TABLE 3.2 Understanding RQNN Parameters

Parameter	Explanation/Indication
β	This is a learning parameter and is thus necessary to update the synaptic weight vector W. For Hebbian learning, it is natural to expect $\beta < 1$
m	This is the mass of the quantum object and is associated with self-excitation of the SWE. To increase self-excitation, keep $m < 1$
ζ	This is the scaling factor to actuate the spatial potential field $V(x)$ and thus causes input excitation (since it appears as a multiplicand in the SWE). The value of ζ can be kept negative or positive
N	This is the number of neurons along the spatial axis. This can be considered as if some quantum object (i.e., each neuron in the form of a window) in the neuronal lattice is looking at the spatial neural network and accepting the input signal sample. Its value depends upon the resolution one wants for the input signal sample (cf. Section 5.4)
n	This is the number of Gaussian functions. It should be chosen to describe the nonlinear map that represents the synaptic connection
g	This is the center of the ith Gaussian function ϕ_i. The range of these centers should cover the range of the noisy input signal. Thus, it should be chosen to cover the complete range of the input signal
$MAXIT$	This is the number of iterative steps that are required for the response of the SWE to reach a steady state for a given computational sampling instant
δt	The SWE equation is converted to the finite difference form with time steps δt
δx	The spatial axis is divided into N mesh points so that x is represented as $x_j = j\delta x$ where j varies from $-\frac{N}{2}$ to $+\frac{N}{2}$

parameters for which the model should perform optimally; some of these have been tuned and some can be kept fixed. The learning rate β (which is necessary to update the synaptic weight vector W), the mass of the quantum object m and the scaling factor ζ (to actuate the spatial potential field $V(x)$) need to be tuned specifically, as the performance is sensitive to these. The parameters that should be kept fixed are the number of neurons on the spatial axis, N, as $N = 50$ or 400, the number of Gaussian kernels ϕ, as $n = 25$. Also, the range of the centers g of these Gaussians should also be selected appropriately (which, as discussed in Section 3.4, should cover the range of the noisy input signal). In addition, the number of iterative steps that are required for the response of the wave equation to reach a steady state to any particular computational sampling instant of the input signal, i.e., the recurrent evolution of the wave packet, should also be selected properly.

Table 3.2 indicates the various parameters and their relevance in the context of the RQNN model.

3.5.3 Filtering Simple Signals

3.5.3.1 Method and Performance Analysis

Before applying the RQNN technique to enhance complex EEG signals, it is sensible to substantiate our claim by implementing the technique to filter simple example signals in the form of DC, staircase DC and sinusoidal signals that have been embedded with a known amount of noise. The RQNN modeling approach should be able to filter out the known amount of noise. The filtering performance is calculated in terms of root mean square error (RMSE) as:

$$\text{RMSE} = \sqrt{\frac{1}{N_s} \sum_k (y(k) - \hat{y}(k))^2} \tag{3.19}$$

where k refers to the signal samples and N_s refers to the total number of signal samples after the RQNN settles down to a steady state.

A DC signal of amplitude value 2 embedded with 20 dB, 6 dB and 0 dB noise is fed to the RQNN model for filtering. Here 250 samples are evaluated within a time interval of 1 s — i.e., the sampling frequency is 250. The parameters of the RQNN model obtained using the Univariate Marginal Distribution Algorithm (UMDA)[8] [323,324] are taken from references [320,325]. The parameters for the DC signal are $\beta = 0.86$, $m = 2.5$ and $\zeta = 2000$. The other parameter values that are kept fixed are $n = 25$ and $N = 400$, and each sample is iterated eight times to stabilize the SWE. Thus, a single sample is recursively iterated eight times before the next sample is fed to the model. The synaptic weights are updated using the learning rule shown in equation (3.18). The value of RMSE for filtering a DC signal is 3.0805e−007, 5.2372e−007 and 3.526e−006 for an embedded noise level of 20 dB, 6 dB and 0 dB respectively. This is displayed in Figure 3.5. A major drawback to this technique is that even with eight iterations over a single sample, the time taken to filter 8 s data is approx. 16 s. Also, when the filter is run for a longer duration (i.e., for a time of 100 s) then the error increases to 0.1 (cf. Figure 3.6 where the RQNN estimated value overshoots the desired DC value from 30 s onwards). In other words, the filter diverges away from its target. These two issues have motivated the development of the revised RQNN technique that is discussed later in the next section.

Figure 3.7 displays the filtering of the DC signal of amplitude 1.7 and emphasizes the importance of the parameter ζ in the filtering process. When the parameter ζ is increased from a value of 900 to 4000 and then

---

[8]Muhlenbein [378] introduced the UMDA as the simplest version of the Estimation of Distribution Algorithms (EDAs) [379] for dynamic optimization problems. More information about the UMDA can also be found in [380].

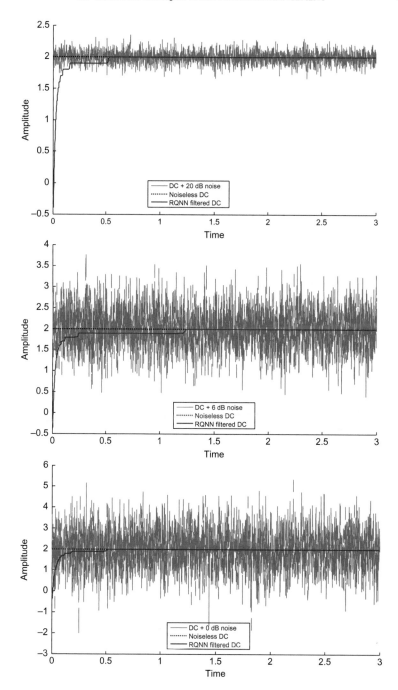

FIGURE 3.5   DC signal filtering (amplitude 2) with the RQNN.

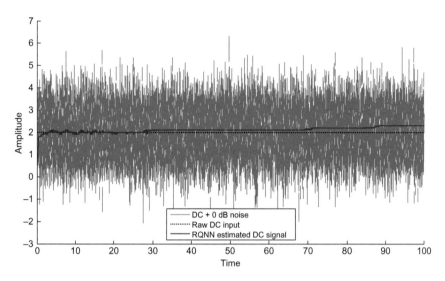

FIGURE 3.6   DC signal filtering with the RQNN over longer duration of 100 s.

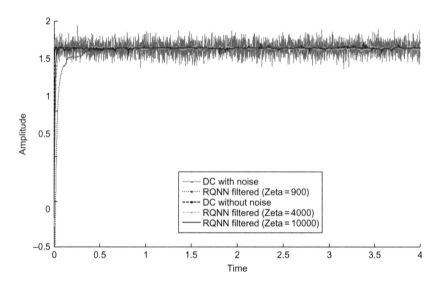

FIGURE 3.7   DC signal filtering with the RQNN.

up to a value of 10000, the response of the RQNN shows a sharp increase, with an increase in the value of the parameter ζ. The rise time (time required for the signal to reach 90% of the final value, which in this example is $0.90*1.73 = 1.53$) increases from a value of 0.0202 to 0.0382 and to 0.1271 when the corresponding value of ζ is decreased from 10000 to 4000 and subsequently to a value of 900 respectively

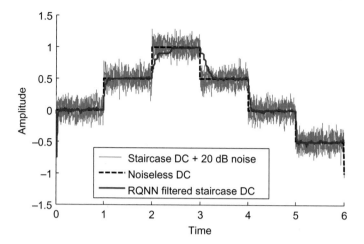

FIGURE 3.8 Filtering a step-wise DC signal.

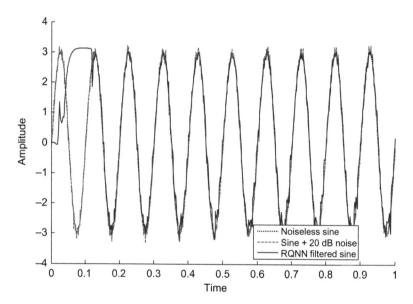

FIGURE 3.9 Sinusoidal signal filtering using the RQNN.

(cf. Figure 3.7). Thus with a lower value of ζ, the response of the RQNN model is sluggish. Figure 3.8 displays the filtering of a step-wise DC signal embedded with 20 dB noise. The result shows that the RQNN is able to filter out the noise and track the time-varying DC signal.

The next example, shown in Figure 3.9, involves filtering a sinusoidal signal of amplitude 3 embedded with 20 dB noise. The parameters of the

RQNN model were kept as $\beta = 0.01$, $m = 1.75$, $\zeta = -250$ and $n = 25$. Here the number of neurons along the x-axis is fixed at 401 (i.e., $N = 400$) and each sample is iterated 75 times before the next sample is fed. The learning rule of equation (3.17) is applied for the weight updating process.

3.5.3.2 Concluding Remarks

These results provide evidence that the RQNN model can filter out the embedded noise within the DC and sinusoidal signals and track the actual signal. The purpose is thus to track the true information within the noisy EEG, in the same way as RQNN has been utilized to track sinusoidal and DC signals in the presence of noise but without the knowledge of the type of noise present [320]. In such situations, the RQNN can also be utilized to enhance the information from raw EEG signals (which consists of noise due to artifact, amplifiers, environment and background EEG effects, etc.). However, the results obtained with this RQNN model for filtering the EEG signals were not satisfactory (cf. Chapter 5). This model is able to reduce noise, but its stability is highly sensitive to model parameters, owing to which, in the case of imperfect tuning, the system fails to track the signal and its output saturates incorrectly. In addition, for different values of DC, the RQNN model requires tuning to a suitable different parameter value [326]. Hence a revision of this model was carried out in order to minimize the instability of the model as well as its parameter dependence. The parameters of the revised RQNN model (as discussed below) can be tuned very easily, and imperfect tuning leads only to under- or over-filtering, without making the system unstable. The next section describes in detail the modifications that have been carried out to this model in order to enhance the overall results.

3.6 REVISED RQNN-BASED SIGNAL ENHANCEMENT

The RQNN model discussed in the previous section utilized the error signal in order to stimulate the neurons in the network and the weights of the network were updated using the two learning rules given in the form of equations (3.17) and (3.18). The revised RQNN model [213,327] presented here has the neurons in the network stimulated directly from the raw EEG signal. In addition, the learning rule for the weight updating process is also different. This revised model is much more stable and the performance is less sensitive to changes in the parameters.

This revised modeling approach uses a complex valued nonlinear SWE to govern the evolution of the quantum wave function $\psi(x, t)$ over the spatio-temporal lattice. The envelope $|\psi|^2$ describes the *pdf* of the signal $y(t)$. The revised approach presented in Figure 3.10 uses a

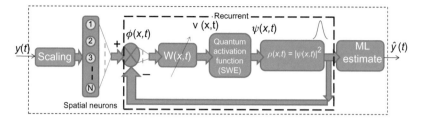

FIGURE 3.10   EEG signal estimation using revised RQNN model. *Reproduced with permission from [327].*

network of neurons with adaptive weights $W(x)$. The received noisy signal $y(t)$ excites this neuronal lattice through its uniformly located centers in the one-dimensional space. The output of this layer is compared with $|\psi|^2$, the estimated *pdf*. The generated error value is weighted by the adaptive weight vector $W(x,t)$ to get the potential function $V(x,t)$.

$$V(x,t) = \zeta W(x,t)\phi(x,t) \tag{3.20}$$

where $\phi(x,t) = e^{\frac{-(y-x)^2}{2\sigma^2}} - |\psi(x,t)|^2$.

This potential modulates the nonlinear SWE, which in natural quantum units is given as:

$$i\partial_t \psi(x,t) = -\frac{1}{2m}\partial_{xx}\psi(x,t) + V(x,t)\psi(x,t) \tag{3.21}$$

The weight vector is adapted so that the envelope $|\psi|^2$ represents the *pdf* of the random process $y(t)$. The estimated value of $y(t)$ is found using MLE as:

$$\hat{y}(t) = E\lceil |\psi(x,t)|^2 \rceil = \int x|\psi(x,t)|^2 dx \tag{3.22}$$

The model can thus be seen as a Gaussian Mixture Model (GMM) estimator of the potential function with fixed centers and variances, and only the synaptic weights $W(x,t)$ are variable. These weights can be trained using any learning rule. In the present model, the adaptation of these synaptic weights is implemented using the explicit learning scheme stated below in equation (3.23).

$$W(x,t+\delta t) = (1-\beta)W(x,t) + \beta\phi(x,t)(1+v(t)^2)\delta t \tag{3.23}$$

i.e.,

$$\frac{\partial W(x,t)}{\partial t} = -\beta_d W(x,t) + \beta\phi(x,t)(1+v(t)^2) \tag{3.24}$$

where the error term is represented as $v(t) = y(t) - \hat{y}(t)$. β is the learning rate and β_d is the de-learning rate. De-learning is used to gradually forget the old information, as the input signal is not stationary, but quasi-stationary in nature. The rightmost term in the above equation is purely positive and so in the absence of the de-learning term, the value of the synaptic weights $W(x, t)$ will keep growing indefinitely. De-learning will thus prevent unbounded increase of these values of the synaptic weights $W(x, t)$ and in the process prevent system instability.

3.6.1 Pseudocode for the Revised RQNN Model

This subsection details the algorithmic pseudocode for implementing the revised RQNN model in Table 3.3.

3.6.2 Understanding the Parameters for the Revised RQNN Model

Similarly to the old architecture, there are three variable parameters that should be tuned properly in order to obtain the desired response of the SWE. These are the learning rate β, the mass of the quantum object m and the scaling factor ζ. However, in this revised model, if the parameters are not optimal, the model does not become unstable but the filtering process is affected, by way of over- or under-filtering. This is because the spatial neurons in the revised model are stimulated directly from the input signal and not from the error signal, as was implemented previously. This model also has another parameter in the form of β_d which is a de-learning factor and primarily used to gradually forget the old information that the model has learned. The number of iterative steps that are required for the response of the wave equation to reach a

TABLE 3.3   Pseudocode for Implementing the Revised RQNN Model

Step	Explanation/Indication
1	Initialize the RQNN after declaring suitable parameter values
2	Feed the next sample $y(t)$ to the RQNN model
3	Calculate the error from $y(t)$ and $\hat{y}(t)$
4	Calculate $\phi(x, t)$ and update the synaptic weight vector $W(x, t)$
5	Calculate the new value of the potential $V(x, t)$
6	Calculate ψ and obtain the *pdf* $\rho(x, t)$ and then estimate (after normalizing the *pdf*) in terms of $\hat{y}(t)$ using MLE
7	Repeat steps 3–6 for a fixed number of iterations
8	Go to Step 2 until all data are to be filtered

steady state to any particular computational sampling instant of the signal should also be selected properly. Also, parameters such as the number of neurons in the spatial axis and the range of the centers of these neurons along with their resolution should also be selected in a suitable way. Thus, the criterion for tuning the parameters is similar to the one discussed for the former RQNN architecture. Table 3.4 indicates only those parameters concerning the revised RQNN model which have not been previously discussed.

3.6.3 Numerical Implementation

This sub-section details the numerical implementation of the revised RQNN modeling approach.

The space variable x is uniformly spaced as $x_n = n\delta x, n = -\frac{N}{2}, \ldots, +\frac{N}{2}$ and the time is spaced as $t_k = k\delta t, k = 1, \ldots, T$. The potential function is approximated as $V(x_n, t_k) = V_n^k$. This potential function excites the nonlinear SWE to obtain the quantum wave function ψ_n^k. Various methods, both explicit and implicit, have been developed for solving the nonlinear SWE numerically, on a finite dimensional subspace [304]. The first approach uses the Crank-Nicolson method [328], which is an implicit scheme for solving the nonlinear SWE and requires a quasi-tridiagonal system of equations to be solved at each step [329]. This scheme, although accurate, requires solving for the inverse of a huge $N \times N$ matrix, which is time-consuming. Hence, the implementation was carried out using the explicit scheme.

$$i\frac{\psi_n^{k+1} - \psi_n^k}{\delta t} = -\frac{\psi_{n+1}^k - 2\psi_n^k + \psi_{n-1}^k}{2m\delta x^2} + V_n^k \psi_n^k \tag{3.25}$$

This method is linearly stable for $\delta t/(\delta x)^2 \leq 1/4$, with a truncation error of the order of $(O(\delta t^2) + O(\delta x^2))$.

Another point to note here is that the normalized character of the *pdf* envelope, $|\psi|^2$ is to be maintained, by normalizing at every step. Thus,

$$\sum_{n=1}^{N} |\psi_n^k|^2 \delta x = 1 \text{ for all } k \tag{3.26}$$

TABLE 3.4   Parameters of the Revised RQNN Model

Parameter	Explanation/Indication
β_d	This is the de-learning parameter and is used to forget the previous information. De-learning is used to prevent unbounded increase of the values of the synaptic weights $W(x, t)$

3.6.4 Filtering Simple Signals

3.6.4.1 Method and Performance Analysis

Before applying the revised RQNN modeling technique for filtering the EEG signals, it is evaluated for simple signals in the form of DC and sinusoidal signals to which a known amount of noise has been added. Thus, the RQNN modeling approach should be able to filter the known amount of noise.

A DC signal of amplitude value 2 is embedded with a noise of 20 dB, 6 dB and 0 dB and fed to the RQNN model for filtering. The number of neurons within the RQNN model along the spatial axis is kept at $N = 400$, and each sample is iterated once to stabilize the SWE; i.e., a single sample is iterated only once before the next sample is fed to the model. The de-learning parameter β_d is kept as 1 unless specifically mentioned. The learning rule for the DC signal is the one displayed in equation (3.23). The parameters of the RQNN can be tuned using any optimization technique. In the present case, the particle swarm optimization (PSO) technique is used. The parameters β and ζ were tuned after fixing the parameter $m = 0.5$. The parameters obtained are $\beta = 0.002$ and $\zeta = 775.05$. Figure 3.11 displays the plots showing the original signal, noisy signals and the RQNN filtered signals. A video showing the movement of the wavepacket for DC filtering is available in reference [330]. The RMSE while filtering the DC signal is 0.00042284, 0.0012071 and 0.0017837 for an embedded noise level of 20 dB, 6 dB and 0 dB, respectively, for the proposed RQNN model which is also shown in Table 3.5. This table is partially reproduced from reference [320] to compare the RMSE of the proposed RQNN filter, Kalman filter and the old RQNN model. It is clear that the performance of the proposed RQNN is significantly better than the Kalman filter [327] and the old linear RQNN for all three noise levels displayed. However, the RMSE of the proposed RQNN is poor compared to the old nonlinear RQNN. An important aspect that needs to be mentioned here is that the old RQNN requires parameter tuning if the amplitude of the DC signal is varied from an amplitude of 2 to an amplitude of 10 [326]. Thus, the stability of this model is sensitive to the values of the model parameters. Similarly, the Kalman filter also requires re-tuning of the model parameters if the amplitude of the input signal changes. However, this is not the case with the proposed RQNN model. Figure 3.12 displays the filtering process with the same parameters for the proposed RQNN model but for DC signals of amplitude 5, 8 and 10 embedded with 0 dB noise. The result shows that the proposed RQNN is able to effectively filter and track the time-varying DC signal. In addition, when filtering the signal for a longer duration of time with the former RQNN model (cf. Figure 3.6), the filtered signal loses its track while the revised model is very stable (cf. Figure 3.12 and Figure 3.13).

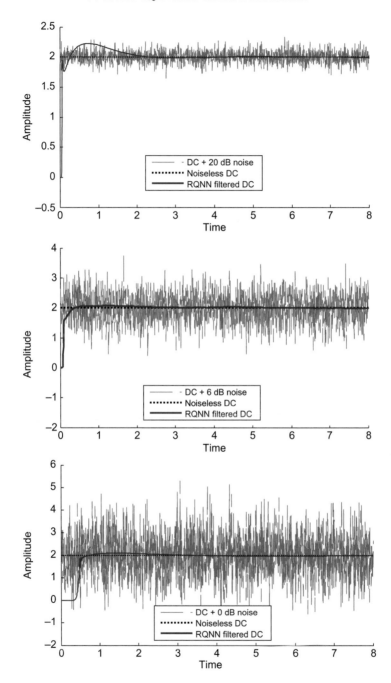

FIGURE 3.11   Filtering DC signal of amplitude 2 embedded with different amount of noise.

TABLE 3.5   Performance Comparison for DC Signal of Amplitude 2

SNR	Kalman filter RMSE	Old linear RQNN RMSE	Old nonlinear RQNN RMSE	Proposed RQNN RMSE
20 dB	0.015	0.014	0.000040	0.0004228
6 dB	0.037	0.060	0.000062	0.0012071
0 dB	0.090	0.150	0.000090	0.0017837

Partially reproduced with permission from [327].

Another difference between the two RQNNs is the number of times a sample is iterated for the evolution within the SWE loop. In the former RQNN model, this value was eight and it has been reduced to just one in the revised model. This saves computational time in the filtering process (the previous model took 16 s while the revised model takes 0.5 s for filtering 8 s of data), although the difference in the value of error between the two models is marginal. Thus, there is a significant improvement in the RMSE in comparison to the previous RQNN model as well as the Kalman filter.

Another example is the tracking of a staircase DC signal with an embedded noise of 20 dB. It is clear from the plot shown in Figure 3.14 that the RQNN is able to track the staircase DC with reasonably good accuracy. The next step is to filter a sinusoidal signal of amplitude 3 embedded with 20 dB noise. The parameters of the RQNN model were kept as $\beta = 5.25$, $m = 0.25$ and $\zeta = 1.75$ and $N = 140$ and each sample is iterated 60 times before the next sample is fed to the RQNN. This is shown in Figure 3.15.

The results show that the proposed RQNN model is able to effectively filter and track a time-varying DC signal, a step-wise DC signal and a sinusoidal signal. The RQNN technique efficiently filters the input signal for a longer duration of time as well as when the amplitude of the input signal is varied (cf. Figure 3.12 and Figure 3.13). In addition, the proposed model is computationally efficient and there is a noteworthy improvement in the RMSE in comparison to the Kalman filter, as is evident in Table 3.5. Thus, the proposed model can now be used to filter EEG signals (cf. Chapter 5 and Chapter 6).

3.7 DISCUSSION

As can be seen from the results of the filtering approach with the RQNN models, there are many parameters that should be selected/tuned appropriately to obtain the desired response of the SWE.

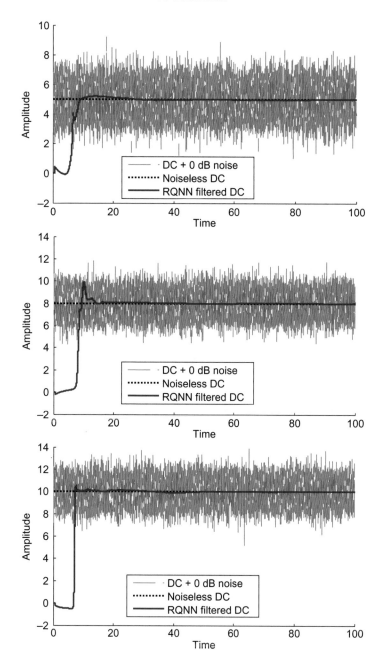

FIGURE 3.12   Filtering DC signal with different amplitude values embedded with 0 dB noise.

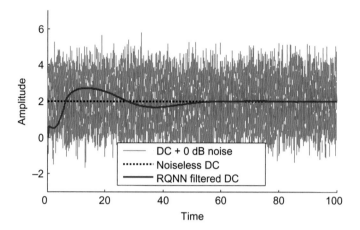

FIGURE 3.13   DC signal filtering with the revised RQNN over longer duration of 100 s. *Reproduced with permission from [327].*

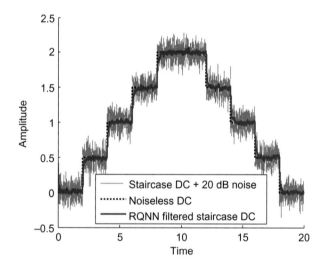

FIGURE 3.14   Staircase DC signal filtering with the revised RQNN model. *Reproduced with permission from [327].*

Selecting the best combination of parameters (cf. Table 3.2 and Table 3.4) is a complex task. The traditional RQNN model requires tuning of suitable different parameter values for different values of DC signals, and imperfect tuning leads to saturation of the RQNN model output to incorrect values (cf. Section 3.5). The revised model is stimulated directly using the raw input signal and is less sensitive to changes

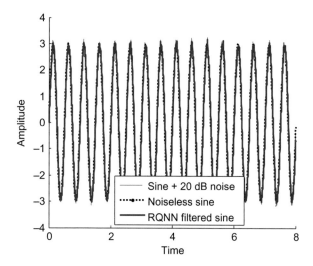

FIGURE 3.15   Sine signal filtering with the revised RQNN model.

in the model parameters. The input signal fed to the RQNN model can be scaled so that the number of neurons required to envelop the noisy input signal is reduced. This further results in reducing the computational load on the movement of the wave packet that tracks the noisy input signal by way of reducing the number of iterations within each computational sample (cf. Chapter 5 for further details on this). The remaining parameters of the RQNN model can be selected using optimization algorithms such as Particle Swarm Optimization (PSO) or Genetic Algorithm (GA) (cf. Section 3.6) and the results from the RQNN approach have shown improvements over the Kalman filter by way of improving the RMSE (cf. Table 3.5).

These points outline the potential advantage of the RQNN framework for filtering noisy signals, and it should be investigated further for filtering real time complex signals such as the EEG.

3.8 CONCLUSION

This chapter has focused on the fundamentals of QM and the application of these fundamentals in practice, using the SWE. The theoretical concepts in applying these QM fundamentals in two different RQNN models have been examined in detail. The RQNN-based techniques have been shown to be effective at filtering noise from simple signals in the form of DC and sinusoidal waveforms, even without changing the parameters of the model. The advantage of these RQNN-based techniques is that they are purely data-driven, are generic and can be applied with little or almost no *a priori* knowledge about the signal that is to be filtered. It is argued that although filters based on neural networks also work by aggregating to provide a coherent representation of the true state (from the different neurons fired in a distributed manner) but these have an underlying mechanistic feature which makes them insufficiently intelligent [331,332]. The RQNN-based filters do not require any kind of *a priori* information about the signal or the noise, and the parameters of the model can be generalized across different amplitude values of the signals. Therefore, the RQNN-based filters may be categorized as sufficiently autonomously adaptive. The pseudocode, the numerical implementation and the importance of each parameter within the RQNN models have been discussed in this chapter.

The main focus in this chapter has been the implementation of QM-based fundamentals in the form of novel RQNN models as filtering techniques. Chapter 5 discusses the experimental results obtained by using the RQNN technique for enhancing nonlinear and non-stationary EEG signals using different learning algorithms [36,37,213].

The Proposed Graphical User Interface (GUI)

4.1 INTRODUCTION

The front-end display of the brain–computer interface (BCI) is a graphical user interface (GUI) user, and this plays a very important role in enhancing the performance of the complete BCI system (cf. Section 2.10.1). This interface can also be referred to as a brain–robot interface for a robot control application. A detailed investigation into the necessity for an interface, the expectations of a good interface design, as well as the designs implemented by various researchers and highlights of their experiences are given in Chapter 2. The major problem with some of these existing interfaces, which were designed for robotic control tasks, is that they aim to control the device with only either a left hand or a right hand movement. A device controlled by such an interface is expected to move forward irrespective of whether the BCI user issues a command or not. In addition, these approaches necessitate some other provisions for halting the device [333, 334]. In order to address these issues, the interface presented in this book looks beyond the left hand and right hand maneuvering tasks. The proposed interface is concerned with the practical requirements of a BCI user, i.e., realistic control of the movement of a device. This requirement needs robot movement to be controlled in any specific direction (say forward, left, right, backward, halt, etc.), and also for destination-specific commands (say Room 1, Kitchen, etc.) to be followed. In addition to these, robot arm pick-and-place[1] tasks should be

---

[1] In the present work, the pick-and-place task refers to the robot arm touching a specific object rather than actually picking it up. The robot arm returns back to its original position after touching the specified object chosen by the BCI user.

carried out by simply controlling the left hand and right hand (two-class) motor imagery electroencephalogram (MI EEG).

The design of the proposed GUI incorporates dynamic and real time behavior by using a shared control approach (discussed in Section 2.11.1) with the aim of developing the following aspects:

1. A two-class BCI can be implemented using any combination of right hand or left hand imagination. In the proposed interface design, which is based on two motor imaginations, both of these imageries are capable of making selection of a choice of any 8 possible commands. A no mental imagination results in no selection of a choice and thereby allows transfer of control to the next two choices.

2. The proposed GUI motivates the BCI user (discussed in Section 2.10.1) to estimate the time of display of the next available choice and thus be prepared to start the MI in advance. However, the use of this approach should depend on the capabilities of the BCI user in controlling his/her MI EEG, which would be determined by careful offline analysis.

3. The proposed GUI updates itself in real time by acquiring information from the environment surrounding the device (mobile robot and robot arm), thereby reducing the number of human—machine interactions, which will in turn result in reduced task completion time. The task completion time is also referred to as mission time.

This chapter will first give a conceptual description of the role of the proposed interface within the BCI system (Section 4.2). This section will also discuss the proposed interface in detail within the BCI framework, as well as explaining how the mobile device can be maneuvered within a changing environment. A flowchart explaining the adaptive behavior of the GUI by using the shared control approach (cf. Section 2.11) is also given here. The section then describes the autonomous interface for mobile robot control applications as well as for robot arm control applications. This work requires software modules that have been written in different programming languages as is described in Section 4.3. MATLAB is used for signal processing issues of BCI[2]. The user interface is designed in Visual Basic (VB)[3], and the object detection and mobility control algorithms for robot control are written

---

[2]MATLAB routines are developed by professional programmers and are highly optimized. Hence, algorithms, analysis of data and experimentation are often implemented in high-level languages like MATLAB.

[3]VB is an integrated, interactive development environment that is highly optimized for developing GUIs for rapid application development.

in C++[4]. This section describes the coordination and the interfacing requirement between these modules. Section 4.4 concludes the chapter.

4.2 OVERVIEW OF THE PROPOSED GUI WITHIN THE BCI FRAMEWORK

An illustration of the proposed interface for a supervised robot control task within the BCI framework is given in Figure 4.1. Before exploring the details of this interface, it is important to understand the position of the proposed interface within the BCI system. Signal processing usually consists of the signal acquisition block, the pre-processing block (implemented using the proposed recurrent quantum neural network [RQNN] model), the feature extraction block (implemented using Hjorth and Bandpower features), the classifier and the post-processing block (implemented using multiple thresholding). This signal processing block continuously sends the class information (which is indicative of imagery of the Left hand, Right hand and No-control [NC] state) to the interface

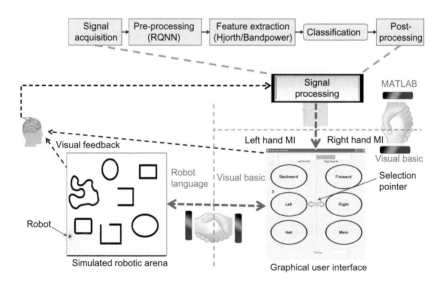

FIGURE 4.1   Proposed interface design within the complete BCI framework.

[4]C and C++ are the languages of choice for hardware interfaces, as they allow access to the memory of a computer or microprocessor to manipulate bits and bytes.

through the MATLAB S-function[5]. The device (any mobile robot/wheel-chair) is simulated in the visual scene of the arena (in Player-Stage [335]) and practically implemented in the Pioneer and Schunk robots[6] [336]. However, controlling the device is only possible through the user interface. The user has to perform either of the two tasks; namely imagining a movement of the right hand or the left hand/foot, but this needs to be synchronized with the movement of the selection pointer arrow on the interface (discussed in detail in Section 4.2.1). The selection pointer arrow allows the BCI user to know in advance the association of his/her MI with a specific control output, say for example, the association between the right hand MI and either Right, Forward or Backward movements is shown in advance on the interface. The control signal from the interface is sent in the form of a final command to the controlled device through the User Datagram Protocol (UDP) [337]. The controlled robotic device receives this signal and acts accordingly, but in agreement with its obstacle or collision avoidance system (whichever is applicable depending on whether the robot is controlled by autonomous or supervised means).

Figure 4.2 displays the GUI in its complete form. This may appear complex at first glance, but it is actually very simple and uniform because any intended task can be selected by mentally imagining moving either the left or the right hand. The purpose of displaying the complete form of the GUI is to show that the topology of the interface remains the same for both the arm control (cf. Figure 4.2(c) and (e)) and the mobility control (cf. Figure 4.2(b) and (d)) tasks. In addition, irrespective of whether the task is supervised or unsupervised/autonomous control, the topology remains uniform, i.e., any task can be selected by performing either of the two MIs. This is an important benefit of the proposed design, and it is intended to make the BCI user feel comfortable in any of the task modes. The BCI user can select any of the control tasks for mobility control (autonomous or supervised)[7] or arm

---

[5]S-functions are dynamically linked subroutines that the MATLAB interpreter can automatically load/execute and hence these provide a powerful mechanism for extending the capabilities of the simulink environment in MATLAB.

[6]The Pioneer P3-DX [381] and Schunk [336] are professional research robots in the Robotics laboratory of the ISRC. The P3-DX robot is equipped with motors, encoders, wheels and 16 ultrasonic sonar sensors. The modular Schunk robot arm has actuators, manipulator and grippers.

[7]The interface for autonomous mobility control provides destination-specific commands such as 'Room1,' 'Room2' etc. while the interface for supervised mobility control provides specific maneuvering commands such as 'Forward', 'Right' etc.

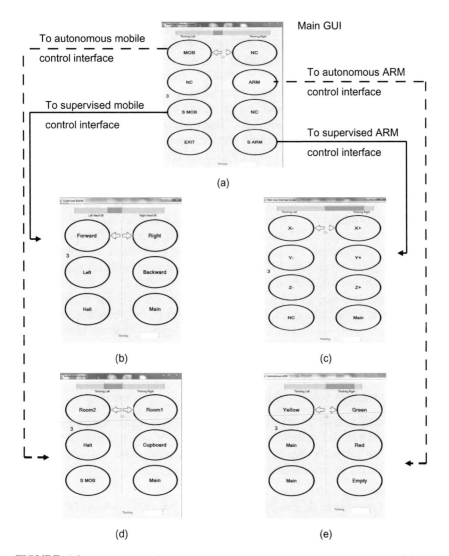

FIGURE 4.2   Framework of the complete graphical user interface. Perform left/right hand MI when the horizontal arrow points to the two options. If the user does not perform MI in the trial time of say 5 s, i.e., no-control indicated by NC, then the horizontal bar moves down to the next two available options. A common interface is designed to issue commands to the mobile robot/wheelchair for autonomous control (MOB) and supervised control (SMOB) as well as for the ARM control (object pick-and-place task). In the MOB interface, i.e., in autonomous mode, the BCI user can issue commands to the robot in terms of predefined locations. With the SMOB interface, i.e., in supervised mode, the BCI user can issue specific commands to move the robot in specific directions.

control (autonomous or supervised)[8] from the main interface, and can traverse between any of these sub-interfaces. The main interface indicates the autonomous mobile control sub-interface as MOB, the supervised mobility control as SMOB, autonomous arm control as ARM, supervised arm control as SARM and no control as NC. The selection of these choices and the proposed sub-interfaces are discussed in detail in the following sections. The purpose of displaying the complete form of the interface in this section is to acquaint the user with the way that the proposed GUI can be utilized for supervised and unsupervised controls of both the robotic arm and the wheelchair/mobile robot. In addition, the topology of any of the proposed interfaces will remain the same for arm and mobile robot control tasks. Thus, the main interface is solely utilized for transferring control to either the mobility control or the arm control interface in response to one of the two mental imageries performed by the BCI user. The first three options in the main interface display an NC choice, as it is easier to differentiate a left vs. NC, right vs. NC than the right vs. left MI.

The next two subsections discuss interface designs for mobility and arm control applications. However, the design of the interface should be understood first, so that the common components can be identified in all the mobility control applications.

4.2.1 Interface for the Mobility Control Application

The mobility control interface in its simplest form is referred to as the non-adaptive/fixed interface. The topology of all the interfaces presented here remains the same, so understanding the working of one design will aid in understanding the remainder.

4.2.1.1 Supervised Mobility Control Interface (Non-Adaptive Form)

Figure 4.3 shows the user interface for supervised control of the wheelchair/mobile robot application. The task is to move the robot/wheelchair in one possible direction, referred to as Left, Right, Forward or Backward or to Halt/Stop, if this (as discussed in Section 2.10.1) is helpful for subjects for whom the trial time has been reduced by cutting down the rest period or the period between subsequent trials. Thus, in these cases, after the user has issued a command, the time available for issuing the next command is reduced. This helps to increase the number

---

[8]The interface for autonomous arm control provides target choice specific commands such as 'Green Ball,' 'Red Ball' (for pick-and-place tasks), while the interface for supervised arm control provides specific maneuvering commands such as 'X+ ' (i.e., move the arm in positive X-axis), 'Y+ ' (i.e., move the arm in positive Y-axis), etc.

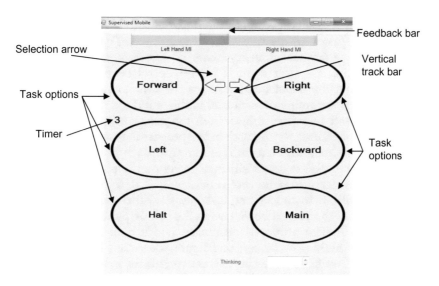

FIGURE 4.3 Proposed user interface. The selection arrow indicates the two available options to the user during any trial. If the arrow points to the options 'Forward' and 'Right', the user can issue the 'Forward' command by performing left hand MI. The vertical track bar and the timer indicate to the user when the selection arrow will move from one set of options to the next. The feedback bar gives sensorimotor rhythm (SMR) feedback.

of possible commands issued within a given time window, i.e., the information transfer rate (ITR). The actual command to drive the robot is sent at an exact time-instant that is appropriate and convenient to the BCI user (usually at the end of the 6th [or 4th] second, which is when the peak offline accuracy for a 7 s [or 5 s] trial is detected after offline analysis of the complete session). At every t_s seconds (i.e., after a complete trial duration) the selection arrow points to the next two available choices. Thus, if the user intends to select the nth choice, then a wait of a system scan time of $t_0 = (n - 1)t_s + t_s = nt_s$ is necessary, where t_s is the scan time period between subsequent choices. If the user does not perform any MI during a scan period then a no-control (NC) state is assumed to be the choice of the user. Therefore, the selection arrow moves down to the next selection option. Thus the user now has to perform one of the two mental imageries but these are now used to choose between Left movement or Backward movement with reference to the robotic device. Again, if the user does not perform either MI, the selection pointer arrow moves to the next available option set. If the value of t_0 exceeds T (the scan time for one complete scan cycle), then it means that the user failed to select the appropriate task in the first round of scanning. Thus, the user has to wait until the selection arrow again points to the desired task. The time required to select the first of the two options is t_s seconds whereas the maximum time required to select the last two options in the first scan cycle is $3 \times t_s$. The interface incorporates voice guidance at every stage of the task selection process, thereby enabling the user to have better control over the selection. A 'beep' sound is the usual indication for the user to start MI, and this happens at the beginning of each trial in this interface.

The proposed interface utilizes both the motor imageries optimally. The number of choices for the user in the proposed design is six or eight, and these have been divided into two groups. Each imagery (either left or right) can be used to select either of the two options in each decision trial, which differs from the conventional BCI designs in which one imagery is utilized to choose one task and the other imagery is used to pass the control to the next choice (as in the hex-o-spell design discussed in Section 2.10). The proposed design, in which both motor imageries are capable of making a choice, considerably decreases the overall number of communications necessary. Moreover, all the available choices of selection are visible to the BCI user at all times and the time at which the selection arrow will point to a particular task option is also known in advance. This can be utilized to further reduce the speed of translation t_s or the trial time of the selection arrow, although this can only be implemented after suitable experimentation. These features of the interface make it very user-friendly and user-centered.

Now attention should be shifted from the simplicity of the fixed nature of the interface in order to appreciate the complexities involved in incorporating the adaptation features. The next subsection details the implementation of the concept of adaptability within the supervised interface.

4.2.1.2 Supervised Mobility Control Interface (Adaptive Form)

The supervised interface for mobile robot control using adaptive behavior is also referred as an intelligent Adaptive User Interface (iAUI), and it has been designed and implemented by utilizing information from the sonar sensors attached to the mobile robot. The iAUI is discussed in three subsections, beginning with the internal framework architecture, a flowchart and an example.

4.2.1.2.1 THE iAUI ARCHITECTURE

The iAUI is composed of four main modules: the communication module (CM), the information refresh module (IRM), the adaptation module (AM) and the monitor module (MM) (which is the front view of the iAUI), as displayed in Figure 4.4.

The first and the most important module is the CM that communicates bi-directionally with the robotic device (receives sonar sensor

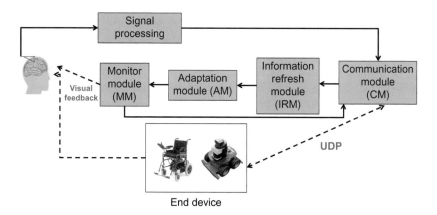

End device

FIGURE 4.4   Framework of the intelligent Adaptive User Interface (iAUI). CM: Communication point between the BCI user and the controlled device. IRM: Gathers information from the robotic device (through the CM). AM: Makes decision with the information from the IRM. MM: Front-end display that alters as per the directions of the AM.

values and issues commands) and uni-directionally with the BCI user (receives post-processed classifier output). The second module is the IRM, which gathers information about the surrounding environment (through the CM). The IRM refreshes the information and interacts with the adaptation mechanism of the AM. The AM, in charge of retaining the existing rules or modifying the same, is responsible for the final adaptability of the MM. The adaptability of the MM refers to the process of modifying the front view of the BCI user screen after the BCI user issues a command or a complete scan cycle of the interface is completed without the user issuing any command. This means that the commands that are offered to the BCI user (i.e., the commands Backward, Forward, Left, Right, Halt and Main Interface as shown in Figure 4.3) will be displayed on the MM such that the most likely command is placed in the topmost location. The two options in the topmost location have the highest probability of being chosen by the BCI user. Therefore, the most likely options are the easiest and quickest to access (cf. Figure 4.7), thereby reducing the decision-making time. The decision-making mechanism of the AM is capable of triggering adaptations of the MM and hence it should be handled with utmost care. These adaptations solely depend upon the dynamic situation of the robot with reference to the environment and do not take user characteristics or preferences into account. Therefore, the adaptive nature of the interface is applicable only during run-time and cannot be initiated solely by the interface components as they do not have any knowledge of the changing characteristics of the dynamic environment.

4.2.1.2.2 FLOWCHART AND STATE MACHINE DIAGRAM

Figure 4.5 is a flowchart of the way the interface works [338]. It shows the control flow of the interface with a typical trial time of 8 s. The command is issued by the BCI user at the end of second 6 and the controlled device issues the interface update feedback (through the CM) after a time interval of a further 1 s or 2 s (depending on the issued command). This small time interval is required because it is expected that the environment will change while the mobile device performs the commanded operation. Therefore, this time interval is taken as an added measure to further enhance the possibility of passing on information about the change in the dynamic environment to the user interface. The interface is allowed to adapt and update only in two cases; first, when the user sends a command to the device and second, when the entire interface scan cycle is completed without the BCI user issuing any command. The purpose of updating the interface when a command is issued has been explained above. The purpose of updating the interface after a complete scan cycle is to incorporate any changes that might have occurred within the dynamic environment during the complete scan cycle of $3 \times t_s$ i.e., approx. 24 s for a typical trial time of 8 s with three choice options.

The protocol for maneuvering the robot/wheelchair can be modeled in the form of a state machine as shown in Figure 4.6. The robotic device to be maneuvered can accept commands while being in motion or at halt. Thus, the device is always in a waiting state to accept commands. The user performs an MI in accordance with the interface and sends a command to drive or halt the robot/wheelchair. During the robotic movement, the user can continue to perform the MI or remain in an NC state as per the requirement of controlling the device. This is an important benefit of the interface: the user can remain in an NC state or the user can issue a command while the robotic device is in motion.

Consider an example, to explain all these critical issues further.

4.2.1.2.3 iAUI OPERATION IN AN EXAMPLE SCENARIO

Figure 4.7 shows the interface in a series of example situations, to help to explain the adaptive nature of the interface. Various adaptive forms of the interface are displayed to show the adaptation process (of the interface). The various modules of the iAUI are listed again:

CM: Communication point between the BCI user and the controlled
    device
IRM: Gathers information from the robotic device (through the CM)
AM: Makes decision with the information from the IRM
MM: Front-end display that alters as per the directions of the AM

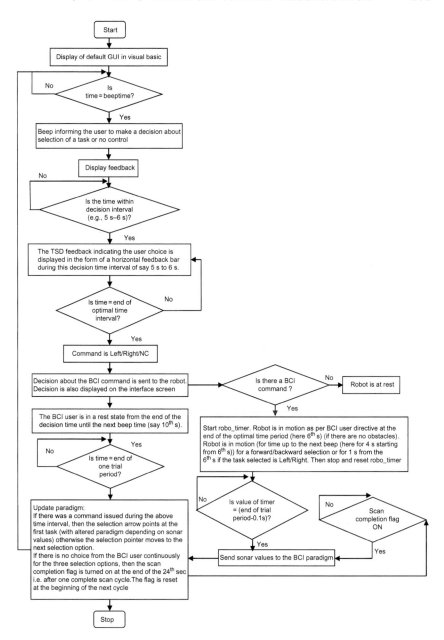

FIGURE 4.5   Flowchart of the user interface.

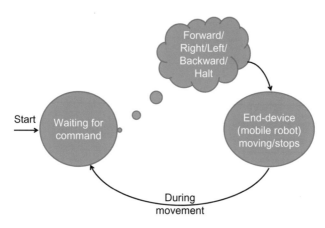

FIGURE 4.6   State machine diagram for device (mobile robot) movement control.

At different locations, the CM receives vital information regarding the obstacles around the robot from the sonar sensors, and these help to update the rules of the AM through the IRM. This helps to update the MM appropriately, and thereby introduces the possibility of displaying the most probable commands expected from the BCI user.

To appreciate the adaptability concept of the interface, assume that the robot begins from position (a) in the diagram shown in Figure 4.7 and is to be maneuvered so that it reaches the position shown by the orange marker in (a)–(h). The arena also contains various obstacles. These are shown in the form of cluttered images. The location of the robot in the arena is displayed on the right hand side of Figures 4.7(a)–(h), and the GUI awaiting the user's mental commands is displayed on the left. At position (a), the two most probable choices displayed are 'Forward' and 'Right.' The BCI user performs a left hand MI and issues the command 'Forward.' When the robot starts to move in the forward direction, the left and right hand sides of the robot get blocked and only the front and backward sides remain open (shown in (b)). This information is sent to the interface in the form of sonar sensor values. Thus, the interface adapts immediately after the user's 'Forward' command and the two most probable choices become 'Backward' and 'Forward.' The user now performs no MI for the complete scan cycle and so the robot continues its forward motion. However, the user had the opportunity to select the 'Forward' and the 'Backward' choices, as these were the most easily available options. Whenever the robot sees an opening, as on the right hand side in (c) (which will be referred to subsequently as branching[9]), it stops and waits for the

[9]The concept of branching is very useful for the BCI user, as the robot always stops whenever a decision is required at a crossroads.

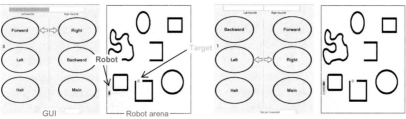

(a) Forward & Right have higher priority than the Backward & Left. Forward command is sent by BCI user through a left hand MI.

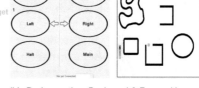

(b) During motion, Backward & Forward have better accessibility, and hence they are prioritized and re-positioned accordingly.

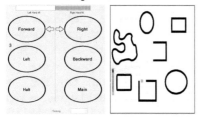

(c) The interface gets updated after completing one scan cycle & hence prioritizes Forward & Right.

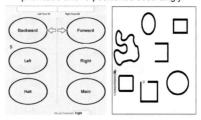

(d) While Right is being honored, Forward & Backward become more accessible; hence, they are again prioritized.

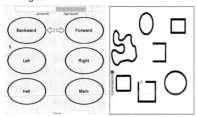

(e) When the BCI user issues Forward, the interface gets updated to (f) as during motion, the right side path gets blocked.

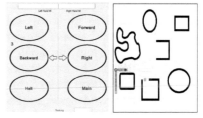

(f) After Forward command, the interface updates by prioritizing Forward and Left as they become more accessible.

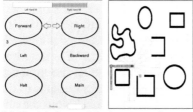

(g) The interface updates here as another scan cycle finishes without the user issuing a command. The user now issues Right.

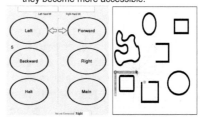

(h) When the rotate Right command is being honored, the left side path becomes more accessible and hence is prioritized.

FIGURE 4.7   Operational example to explain the adaptable nature of the iAUI.

BCI user to issue a command. Simultaneously, in the present situation, the interface has completed its full scan cycle without receiving any command from the BCI user, therefore when the new scan cycle begins (as in (c)), the interface is expected to update using the information received from the sonar sensors. Thus, at the end of the completion of this full scan cycle, the interface receives information from the robot sonar sensors suggesting three probable openings for movement; forward, backward and right. The interface has rules within the AM that give higher priority to forward movement and subsequently to the right, left and the backward movements. The backward movement is assigned the lowest priority, because it is assumed that this is the least likely choice. Therefore, even if the backward option is available, it will not be in the first two choices available to the user unless it is the only available choice. Hence, interface in (c) lists the two most probable options as 'Right' and 'Forward.' The user now performs a right hand MI and selects the task 'Right' within the first trial itself. Thus, the user was not required to wait. When the robot begins its right hand movement, both the right and left hand sides get blocked and this information is passed on to the AM. Hence, while the interface is executing this right hand movement command, the 'Forward' and 'Backward' choices become more probable and are displayed on the interface as shown in (d) and (e). This updating step is possible because the interface is updated 1 s after it has sent the command to the mobile robot, and as discussed earlier, during this small time interval the surrounding environment has changed. The information about the changes in the surrounding environment has been passed to the interface so that the MM can be updated. The user now performs a right hand MI to select 'Forward' (see part (e)). Again, when the robot begins its forward movement, the right hand side path gets blocked (see (f)) and hence the interface now provides the first two probable choices to the user as 'Left' and 'Forward.' But assume the user now does not perform any MI until the robot stops again at the next branch (see part (g)), where the right hand path gets unblocked and because of its higher priority, the interface now updates again to display the first two probable choices as 'Right' and 'Forward.' The user now issues the command 'Right' with a right hand MI. However, the right side again gets blocked (see part (h)) and so the interface again updates to display the two most probable options as 'Left' and 'Forward.' Now the user can issue a 'Forward' command by a right hand MI and reach the target destination.

This example demonstrates the complexities involved in the communication process between the interface and the mobile robot device for appropriately updating the interface. However, since adaptivity has been incorporated into the iAUI, it can adapt itself appropriately in real time, and display the interface in the varying forms shown in Figure 4.7 in order to receive the most probable command from the BCI user. Had there been no adaptability in the interface, the user would have had to issue an NC command in order to reach the second choice option in a static interface and then issue the command to rotate 'Right.' The inclusion of adaptive characteristics in the interface allows the BCI user to drive the robot by issuing the appropriate command in the first instance of the scan cycle rather than issuing two commands from a static interface by issuing an NC command. This strategy saves the BCI user issuing one more NC command and a period of time equivalent to one trial time.

Another major advantage of the adaptive interface is that even if a specific task (say, 'Backward') is the least probable, it is still available to the user at all stages. The purpose of this is to give the user complete control, with all the options available (but prioritized) rather than removing some choices just because the mobile robot device does not expect that command. Thus, this interface is designed by making the user the top priority – it is a user-centric design.

This adaptive approach as well as the non-adaptive interface approach is compared with the existing interface designs by calculating the overall task completion/mission time and the cost incurred for the same in Chapter 6. The next sub-section discusses the interface in an autonomous mode for the mobile robot control task.

4.2.1.3 *Autonomous Mobility Control Interface (MOB)*

Figure 4.2(d) displays the GUI for autonomous control of the mobile robot (referred to as MOB). This interface is non-adaptive and therefore fixed. The user is required to issue commands to the robot in terms of specific locations/destinations, so that the robot can reach these destinations using the predefined map stored within it, but by autonomous means. Thus, the BCI user can select a particular destination, which can be the kitchen, shower, Room1 etc., from the interface by associating their MI with the position of the selection pointer. The choices displayed on the interface and their selection follow the approach discussed in the previous section; i.e., any task can be selected by using a two-class MI. The mobile robot can reach the destination specified by the BCI user by utilizing an inbuilt obstacle avoidance technique. Figure 4.8 displays an example to help to explain the autonomous interface. The user is required to perform one right hand MI to select the destination marked 'Room1.' The purpose of this interface is intended to guide the BCI user to reach the doorstep of this particular room, or the location marked in the figure. Once the robot reaches the destination, i.e., the entrance to 'Room1,' through autonomous navigation, smaller-scale maneuvering/control can be implemented at the discretion of the BCI user through the commands within the supervised iAUI mode. Thus, the interface is designed to provide true independence to the BCI user, by maneuvering between

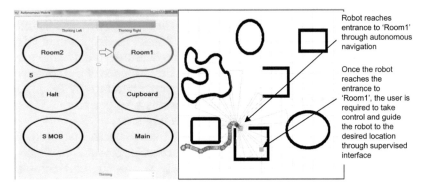

FIGURE 4.8   Mobile robot control through an autonomous (MOB) interface. The user is required to perform one right hand MI through the autonomous interface and select 'Room1.' The mobile robot reaches the entrance to 'Room1' through autonomous navigation. Once this has been done, the user takes control of the robot using the supervised iAUI interface.

locations that are further away more quickly by autonomous navigation, and then using more localized control through a supervised method.

In summary, the controlled device/robot receives the commands from the BCI user through the UDP and navigates either in supervised mode (through the commands Forward, Backward, Left, Right or Idle/Stop) or through the autonomous mode (using location/destination-specific commands). When the device is in motion, an inbuilt collision/obstacle avoidance algorithm avoids collisions and thus maneuvers safely. It is worth mentioning here that the autonomous mobility control mode incorporates obstacle avoidance because the robot knows the location of the destination target from the predefined map. In contrast, the supervised mobility control mode incorporates a collision avoidance scheme because the user issues specific directives and the robot is not aware of the end target destination.

The next subsection discusses the interface for a robot arm pick-and-place task.

4.2.2 Interface for Arm Control Applications

The interfaces displayed in Figure 4.2(c) and (e) are for the supervised and autonomous arm control applications respectively. In the supervised arm control application (SARM), the BCI user is expected to perform the usual left hand or right hand MI (two-class) to command the robot arm movement specifically in the X, Y and Z directions. The resolution of the movement of the robot arm is decided at the robotic end. This is useful to control the robot arm at a very precise location in the robot

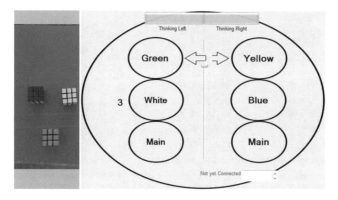

FIGURE 4.9   The camera view and the user interface displayed to the user for ARM control. The object detection algorithm identifies the number and type of objects from the camera view and sends this information to the user interface, which updates itself at the end of a trial or before refreshing the interface. Hence, the three objects identified are displayed at the user interface end.

workspace. However, considering the low accuracy of BCI systems, this mode of operation is unlikely to be the choice of the BCI user.

The autonomous arm control interface (ARM) is displayed in Figure 4.2(e) and Figure 4.9. This interface is expected to allow the BCI user to perform a pick-and-place operation. Whenever the BCI user selects the arm control interface (ARM on the main interface displayed in Figure 4.2(a)), the robot arm sends information about the list of available objects (based on color information) to the user interface. The arm interface shown in Figure 4.9 is displayed, and thus the list of objects is made available for user selection. In addition to this, the live camera view is also displayed on one side of the user screen. Again, object selection can be performed through the usual left hand/right hand MI (two-class mental imagination).

Figure 4.10 shows an example operation, to explain the object selection task. The Schunk robot arm and the three objects are shown in (a), the view of the three objects from the camera is shown in (b) and the ARM interface is shown in (c). The ARM interface displays the list of available objects as identified by the algorithm from the camera view. The user can choose only those objects that the algorithm in the robot is able to identify. The purpose of this limited display of objects is justified by the fact that the robot will be able to complete the BCI commanded task successfully (of object pick up) only if it is able to identify the specified object (through its object-detecting algorithm), otherwise it will fail to complete the BCI issued command. The adaptive strategy approach that is considered within this interface is that whenever the BCI user issues a command, or whenever the complete scan cycle of the interface

  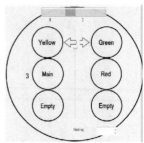

(1) Schunk and 3 objects   (2) Robot camera view   (3) User interface with 3 objects

(a) The algorithm identifies the number and type of objects from the camera view and sends this information to the user interface. As shown here, there are three objects in the camera view and the same gets displayed at the user interface end for selection purpose. The user is performing a left hand MI to select the 'Yellow' object.

  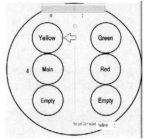

(1) Schunk and 3 objects   (2) User interface with 3 objects

(b) The user has issued the command and selected the 'Yellow' object. When the robot is executing the said operation, the green-colored object is removed manually from the robot camera view.

  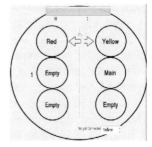

(1) Schunk and 2 objects   (2) Robot camera view   (3) User interface with 2 objects

(c) When the robot returns to its original position, it sends 'Success'/'Failure' information to the user interface. In addition, when the robot returns to its original position, it again sends the information about the number and type of objects available in the camera view to the interface. Since there are now only two objects in the arena, this information is passed on in real-time to the user interface. Therefore, when the user interface gets updated, there are only two objects for selection for the user. The remaining tasks can be marked empty or 'Main'.

FIGURE 4.10 Understanding the user interface for ARM control in a real time scenario.

is completed without the issuance of a command, then the robot sends a new list of objects which it has identified so that the interface is updated in real time. As shown in Figure 4.10(a), the user selects the object marked 'Yellow' by performing a left hand MI. While the robot arm is performing the commanded task, another object marked 'Green' is removed from the camera view as shown in Figure 4.10(b). Therefore, when the robot arm has completed its object pick-and-place task and returns to its original position, there are only two objects in the camera view as shown in Figure 4.10(c). The algorithm at the robot end continuously monitors the objects in the camera view and updates the ARM interface in real time, as shown in Figure 4.10(c).

This section completes the discussion about the proposed interface that has been designed for supervised and autonomous robot control. The next section discusses the important issue of establishing the communication link between MATLAB (which deals with signal processing issues) and VB (which deals with the design of the user interface).

4.3 INTERFACING MATLAB AND VISUAL BASIC

This section gives a brief overview of the approach undertaken to enable communication between the signal processing implemented in MATLAB and the GUI implemented in VB. The S-function in MATLAB plays a very important role in the present work, as it provides a means of communication by passing on the information about the post-processed classifier output, i.e., time-varying signed distance (TSD) from the MATLAB signal processing block to the graphical interface designed in VB through the UDP protocol. UDP is a member of the internet protocol suite, and it uses a simple communication model without explicit hand-shaking dialog by assuming that error checking and correction are not necessary or are already performed in the application. This makes the UDP faster and suitable for a time-critical application such as the one discussed here, where dropping packets is preferable to using delayed packets.

The S-function block in MATLAB continuously sends a post-processed numerical value indicative of either of the two classes or the NC to the interface in VB through the UDP. The 'Listen' thread in VB continuously receives these numerical data (indicating the class to which a sample belongs) in the form of a string. The vertical track arrow in the interface is indicative of the trial time, and this moves continuously to inform the user of the timing in this synchronous GUI design concept. The variables 'countR' and 'countL' in VB count the number of times the classifier output exceeds the upper threshold or the lower

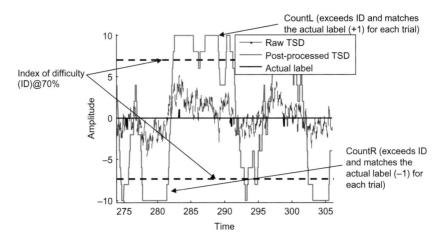

FIGURE 4.11   Post-processed classifier outcome (32 s segment data).

threshold respectively (cf. Figure 2.12) during a 1 s window in the post-processed TSD. This is displayed in Figure 4.11. The final decision regarding task selection is taken in the optimal time interval, which is usually 5 s to 6 s for a trial time of 8 s, and between 3 s and 4 s for a trial time of 5 s. The variable with the larger value, as long as it exceeds the Index of Difficulty, during the optimal time interval is chosen as the winner, and the choice is indicated by selecting the task that the selection arrow indicates. Thus, not only is the BCI output, in terms of a right hand or a left hand MI, taken into consideration from the threshold, but a second parameter (Index of Difficulty) in the form of a confidence level estimation is also considered so that a task gets selected only if the MI exceeds the confidence threshold level. Whenever the user makes a selection, the track bar restarts its continuous movement from the topmost position and the selection arrow returns to the first position. During this time, i.e., when the interface is reset, the MM is updated with information received from the controlled device as discussed in the previous section.

4.4 CONCLUSION

In this chapter, the novel, adaptive, user-friendly, interface design has been discussed. This is an important aspect of the whole system, which enhances overall communication in this time-critical, robot control, BCI application. The chapter has detailed the proposed adaptive interface design, which is user-centric and simple. The dynamic positioning of the choices within the iAUI, where the most likely choices are displayed as the first option, is implemented on the basis of feedback from the controlled device (sonar sensor information is sent from the robot to the interface). The resulting interface design is made simple for the BCI user because it only requires two classes of MI to select from up to six options. In addition, the concept of adaptive shared control has been successfully implemented and discussed in this chapter. Chapter 6 presents the results of the iAUI for a real time robot control task.

The next chapter discusses the quantum mechanically based RQNN technique that has been applied to EEG signal pre-processing for signal enhancement.

Recurrent Quantum Neural Network (RQNN)-Based EEG Enhancement

In Chapter 3 it was shown how the Recurrent Quantum Neural Network (RQNN) framework can be applied to filter signals. This chapter shows its effectiveness in filtering the electroencephalogram (EEG) signals from the datasets of the brain−computer interface (BCI) competition II and IV [339,340]. Each competition dataset consists of EEG signals acquired from nine subjects. There are many parameters in the RQNN model that require subject-specific tuning. This chapter discusses a two-step, inner-outer, five-fold, cross-validation approach that is utilized to select the subject-specific RQNN parameters. It is shown that the performance of the classifier during the training and evaluation stages with the subject-specific RQNN model is enhanced compared to that using raw EEG, Savitzky-Golay (SG) filtered EEG or even raw EEG with the power spectral density or Bispectrum-based features [327]. A statistically significant performance improvement with the RQNN is demonstrated across multiple sessions. In fact, the performance enhancement seen with the RQNN model is also significantly better than those obtained by the winner of the BCI competition discussed in reference [341]. This is one of the major contributions of this book.

5.1 INTRODUCTION

A typical EEG signal is 10 to 100 microvolts when measured from the scalp [30]. Because of the very low signal amplitude, noise has a notable effect on the quality of the measured EEG signals [342]. This noise is embedded within the EEG due to several factors as discussed previously, ranging from 50 Hz power line interference, impedance fluctuation because of loose placement of electrodes, minor body movements/motion artifacts leading to electrode movements over the skin, sweating, electromyogram (EMG)/electrooculogram (EOG) interference, noise introduced due to instrumentation or electronic devices, etc. This section of the book details investigations into use of RQNN models for enhancing the EEG signal. This novel RQNN filtering approach is constructed using a layer of neurons within the neural network framework that recurrently computes a time-varying probability density function (*pdf*) for the measurement of the observed signal. The raw EEG sample is encoded in terms of a particle-like wave packet that recurrently evolves under the influence of the Schrodinger Wave Equation (SWE) and an unsupervised learning scheme, without making any specific assumption about the signal type or the noise. Implementing this quantum mechanically (QM) based conceptual approach in BCI has not been undertaken before.

Since the BCI presented here consists of two channels (C3 and C4) of the EEG data, each channel data is fed through a separate RQNN model to obtain the enhanced signal for the respective channel as shown in Figure 5.1(a). Here, the EEG data from the C3 and C4 channels are fed to two RQNN models, an estimate of the signal is obtained for the samples from both these channels, after which suitable features are extracted from the signals produced by the RQNN. These features are then fed as inputs to train the classifier, which in the present work is the Linear Discriminant Analysis (LDA) [221], Support Vector Machine (SVM) [221], Regression Analysis and another Common Spatial Pattern (CSP) motivated LDA classifier. Once the offline analysis is complete and the classifier has been trained, the parameters/weight vector of the classifier are stored for later use, when the online model is used to identify unlabeled EEG signals during online analysis. In order to capture the time-varying property of the continuous EEG signal, the weight updating process of the RQNN model is continuous (to filter the raw EEG signal) during both the offline and online stages. The system parameters are tuned offline in order to have fixed weighting parameters during the online classification process. In BCI, cross-validation (CV) is carried out to obtain the most stable time-point of the Motor Imagery (MI) performed by the Brain–Computer Interface (BCI) user during the event-related time-period across all the trials of a complete session. In this study, the most stable time-point was obtained using five-fold CV (cf. Section 2.9.1) and hence the parameters of the classifier are chosen at this specific time-point. The performances of both the RQNN models are compared with the raw EEG in Sections 5.2 and 5.3. In Chapter 3, the performance of the RQNN models was compared with the Kalman filter when filtering simple example signals, and the proposed RQNN model was seen to be significantly better. The performance of the

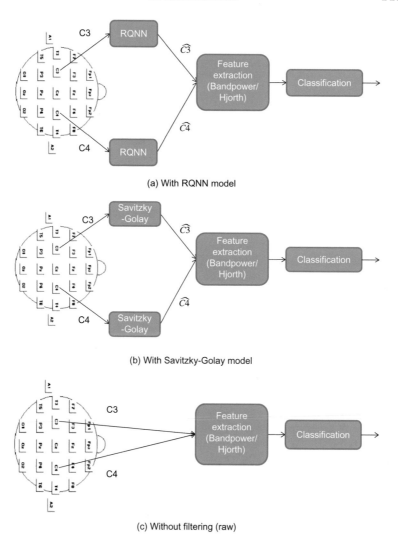

FIGURE 5.1 Framework of the EEG signal filtering for performance evaluation. (a) With RQNN model. (b) With Savitzky-Golay model. (c) Without filtering (raw).

proposed RQNN model is compared with the well-known SG filtering technique and the raw signals while filtering the EEG signals. SG filtering and the raw unfiltered EEG are shown in Figure 5.1(b) and Figure 5.1(c), respectively. During the process of enhancing the EEG signals, the complete set of time series data is fed to the model one sample at a time. Thus, if the signal to be filtered is an EEG time series of 160 trials with an 8 s trial time and a sampling frequency of 250 Hz, then the complete time series of $160 \times 8 \times 250 = 320000$ data points is fed one data point at a time to the RQNN model.

Sections 5.2 and 5.3 discuss the performance of both the RQNN models, beginning with the traditional RQNN model.

5.2 TRADITIONAL RQNN MODEL FOR EEG ENHANCEMENT

This section details the performance of the BCI in terms of accuracy and kappa values while utilizing the traditional RQNN model (discussed in Section 3.5) for evaluating subjects from the BCI competition IV dataset [340]. EEG filtering using the RQNN model is initially investigated with heuristically set parameters and without any scaling of the input signals. Because the input EEG signal is not scaled prior to being fed to the RQNN model, the wave packet is required to envelop a large spatial distance. The approach therefore becomes computationally expensive for online real-time implementation. However, as this laid the foundation for further development of the RQNN models, this section begins with a brief discussion of this approach.

5.2.1 EEG Filtering without Scaling

The traditional RQNN model was first implemented on three subjects from the BCI competition IV2b dataset without any scaling of the input EEG signal [37]. The parameters used to obtain the RQNN filtered EEG signal were set heuristically as $\beta = 1.16$, $m = 0.96$ and $\zeta = 100$, with non-linear modulation of the potential field. β as usual is the learning rate necessary to update the synaptic weight vector, m is the mass of the quantum object and ζ is the scaling factor to actuate the spatial potential field $V(x)$. The number of iterative steps that are required for the response of the wave function to reach a steady state to any particular computational sampling instant of the EEG is 25. Thus, a particular sample is iterated 25 times before the next sample is presented. These parameters were obtained after trial and error over a small set of EEG and kept the same for the three subjects that were investigated with this model. The peak value of the classification accuracy (CA) obtained by using the LDA classifier with the RQNN model (i.e., with the RQNN estimated EEG and the RQNN wave packet signal) and without the RQNN model (i.e., with the features obtained from the raw EEG signal) are indicated in Table 5.1. The value of CA obtained using the LDA classifier with the RQNN filter is better than that obtained using the raw EEG signals by 2.27−5.9% during the training phase for the three

TABLE 5.1   Comparison in Terms of CA Using LDA Classifier

Subjects	Features obtained from raw EEG	Features obtained from RQNN estimated EEG	Features obtained from RQNN wave packet
B0703T	69.10	74.26	75.00
B0803T	87.50	89.42	90.38
B0903T	84.85	85.61	87.12

TABLE 5.2   Comparison in Terms of CA Using SVM Classifier

Subjects	Features obtained from raw EEG	Features obtained from RQNN estimated EEG	Features obtained from RQNN wave packet
B0703T	77.94	82.35	82.35
B0803T	92.00	96.10	100.00
B0903T	90.91	92.42	92.42

subjects investigated. The same is also examined using the SVM classifier, as detailed in Table 5.2. In this case, the value of CA obtained using the RQNN model is better by a margin of 1.51−9.00% when compared with that using the raw EEG. However, these results are obtained on the training dataset to set up the classifier. Once the classifier has been trained, the parameter set obtained is stored for use with test or evaluation data. Using evaluation data, the RQNN approach gave better performance in terms of CA with one subject (B0704E). However, for subjects B0804E and B0904E, the raw EEG performed better in terms of CA than the RQNN approach. However, this work was the first to investigate the RQNN approach for EEG filtering, and this was a good motivation for further work. It is intuitive to apply the same RQNN approach over partially scaled data, as that would mean using a larger number of spatial neurons over a reduced range of the input signal, thereby resulting in a wave packet that could be more effective in enveloping the noisy input signal. This is discussed next.

5.2.2 Scaling the EEG Prior to Filtering

In this part of the study, the input EEG signal during both the training and the evaluation stages is scaled in the range of 0 to 2 and is then fed to the RQNN model. In addition to this, the model is only fed with MI-related data; i.e., data from 3 s to 7 s are fed to the RQNN during both training and evaluation stages. This change is purely to reduce the computational load by not running the EEG enhancement process during the non-event-related period. The three parameters used to obtain the RQNN-filtered EEG signal were set heuristically as $\beta = 0.86$, $m = 1.75$ and $\zeta = 2000$ for the RQNN with nonlinear modulation of the potential field. The number of neurons along the X-axis is fixed as $N = 400$. In addition, the number of iterative steps required for the response of the wave function to reach a steady state at any particular computational sampling instant of the EEG is 50. Thus, each sample is iterated 50 times before the next is presented. The weight of the RQNN model was updated using the Hebbian learning rule [322] (cf. Section 3.5). All these parameters were obtained after suitable trials and experimentation over a small set of EEG data, based on the work reported in Chapter 3.

The enhanced EEG signal is thus obtained by passing the scaled EEG signal through the RQNN model. The next task is to obtain simple Hjorth features and perform signal classification as per the normal procedure using the LDA classifier. Table 5.3 displays the maximum of CA and kappa after the classification process for the RQNN filtered EEG and the raw EEG. Figures 5.2 and 5.3 display the CA and Time-Varying Signed Distance (TSD) plots of the training and evaluation dataset for a representative subject (B04). As is evident from the table, this approach shows better results in the evaluation set, but only for five out of the nine subjects [36]. Thus, amendment of the RQNN model was necessary. In addition to this, the approach for scaling the input signal was also modified. The revised RQNN model explained in Section 3.6 is therefore investigated on the BCI competition datasets with a different scaling approach. This is discussed in the next section.

5.3 REVISED RQNN MODEL FOR EEG SIGNAL ENHANCEMENT

The revised RQNN model was applied to enhance the EEG signals. However, as discussed before, the method of scaling the input EEG signals in the region between an amplitude of value 0 and 2 is carried out a little differently. It is possible to scale the input signal accurately at the training stage because the complete EEG signal from all the channels is available in its entirety. Hence, the signal can be scaled from the minimum and maximum values of the same. However, during the evaluation stage, the whole signal is not available, because it is an online process and the signal is only available one sample at a time. Therefore, during the evaluation stage, the EEG signal is scaled approximately in the range of 0–2 by using the maximum of the amplitude value obtained from the offline training data of that specific channel. The net

TABLE 5.3 CA and Kappa for EEG (scaled) Filtering Using Traditional RQNN Model

Subj.	Training (03 T)				Evaluation (04E)				Evaluation (05E)			
	Max. Acc.		Max. Kappa		Max. Acc.		Max. Kappa		Max. Acc.		Max. Kappa	
	RQNN	Raw	RQNN	Raw	RQNN	Raw	RQNN	Raw	RQNN	Raw	RQNN	Raw
B01	77.50	81.87	0.55	0.64	71.87	71.25	0.43	0.42	61.25	63.75	0.22	0.26
B02	75.00	66.87	0.50	0.34	65.83	60.00	0.32	0.13	61.25	58.12	0.22	0.16
B03	82.50	69.37	0.65	0.39	68.12	63.75	0.36	0.28	85.00	60.00	0.7	0.20
B04	97.50	96.87	0.95	0.94	93.12	91.87	0.86	0.84	79.37	76.25	0.59	0.52
B05	78.75	78.12	0.57	0.56	67.50	89.37	0.35	0.79	70.62	85.00	0.41	0.70
B06	66.25	73.75	0.32	0.47	77.50	66.25	0.55	0.32	80.62	65.62	0.61	0.31
B07	76.25	70.00	0.52	0.40	63.12	63.12	0.26	0.26	63.75	64.38	0.27	0.29
B08	88.13	86.25	0.76	0.72	85.00	86.25	0.70	0.72	93.12	90.00	0.86	0.80
B09	83.75	87.50	0.67	0.75	76.87	86.88	0.54	0.74	70.62	83.12	0.41	0.66
Avg.	80.62	78.95	0.61	0.57	74.32	75.41	0.49	0.50	73.95	71.80	0.48	0.43

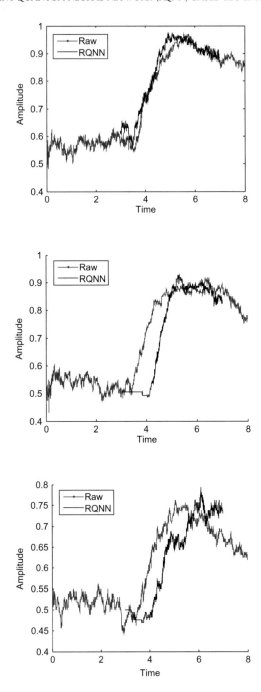

FIGURE 5.2   CA plot for subject B04 on the raw EEG and RQNN estimated EEG signal.

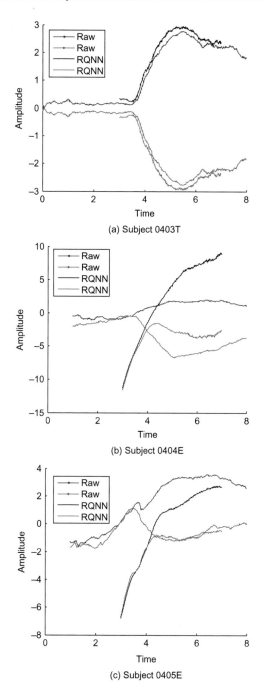

FIGURE 5.3   TSD plot for the subject B04 on the raw EEG and RQNN estimated EEG signal. (a) Subject 0403T. (b) Subject 0404E. (c) Subject 0405E.

effect is that the input signal during the online process is maintained approximately in the region 0–2. The scaled EEG signal may have negative values, or go beyond +2 during the evaluation process. Therefore, the revised RQNN model with scaling was initially implemented with a large number of spatial neurons. Their number was subsequently reduced to decrease the computational load.

5.3.1 Scaling the EEG Prior to Filtering (with a Large Number of Spatial Neurons)

The revised RQNN model, using a large number of spatial neurons, was initially investigated with parameters that were set heuristically. Parameter $\hbar$ was set to unity and the other four parameters were $\beta = 2.7$, $m = 1.75$, $\zeta = 15$ and $\beta_d = 1$, for RQNN with nonlinear modulation of the potential field. The number of neurons along the spatial axis was fixed at $N = 400$. This number is possibly higher than necessary, as they cover an input signal in the range of -20 to $+20$ while the input training dataset of the EEG signal is scaled in the range of 0 to $+2$. However, because this was the beginning of the investigation into merging the concepts of quantum mechanics and EEG with the revised model, this was implemented without calculating the computational expense. The evaluation dataset of the EEG obtained during the online process is unknown and therefore it is scaled approximately by utilizing the maximum of the value of the EEG data obtained from the training set.

The RQNN enhanced signal was only used to obtain the Hjorth features in order to test its suitability for EEG filtering. The maximum CA and the kappa values obtained by only using the Hjorth features are listed in Table 5.4 (3T training dataset), Table 5.5 (4E evaluation dataset) and Table 5.6 (5E evaluation dataset) on the BCI competition IV 2b dataset. For comparison, the raw signal was also scaled in the same way as the signal that is fed to the RQNN. A normal five-fold CV is performed to obtain the time-point with the highest mean kappa coefficient value (average of the five test partitions). Thus, the training set data is partitioned into a training set (80% of the data) and a test set (20% of the data) which is used to calculate the CA rates. The stable classifier parameters at the best time-point from the five-fold CV are utilized to obtain the evaluation stage performance. The results show that the performance of the BCI system is enhanced using this practically feasible scaling approach with the revised RQNN model. The average performance of all the classifiers is consistent in the training and the evaluation stages. Therefore, the next task was to enhance the speed of the RQNN filtering approach by reducing the number of neurons in the spatial axis, to allow the model to be utilized practically in an online, real-time application. This is discussed next.

TABLE 5.4　Training with Five-Fold Cross-Validation (neuronal centers spanning −20 to +20 spatial region)

Subject	LDA		SVM		REG		LD4		LD5	
	RQNN	Raw	RQNN	Raw	RQNN	Raw	RQNN	Raw	RQNN	Raw
B01	81.25	78.13	76.88	79.38	75.63	76.25	76.25	78.75	76.25	77.50
B02	78.75	59.38	75.63	65.63	76.88	63.75	76.25	62.50	76.25	58.13
B03	75.62	66.25	71.88	61.88	68.13	57.50	70.63	60.63	71.25	63.13
B04	99.37	96.25	96.88	95.00	99.38	95.63	99.38	95.00	98.75	95.00
B05	80.62	80.00	79.38	76.88	80.00	76.25	79.38	76.88	78.75	76.88
B06	74.37	69.38	75.63	61.25	75.00	70.00	72.50	63.75	75.00	67.50
B07	88.12	68.13	88.13	66.88	88.75	68.13	88.75	66.25	89.38	68.75
B08	87.50	83.75	89.38	87.50	85.63	84.38	87.50	85.63	87.50	84.38
B09	88.12	86.88	85.63	86.25	86.88	86.88	86.88	86.88	81.88	83.13
Average	83.75	76.46	82.15	75.63	81.81	75.42	81.94	75.14	81.67	74.93

TABLE 5.5 Evaluation 4E with Classifier Parameters Obtained from Training Data with Five-Fold CV (neuronal centers spanning −20 to +20 spatial region)

Subject	LDA		SVM		REG		LD4		LD5	
	RQNN	Raw	RQNN	Raw	RQNN	Raw	RQNN	Raw	RQNN	Raw
B01	73.75	61.25	67.50	58.13	71.88	64.38	70.00	60.00	67.50	60.00
B02	63.33	65.83	64.17	61.67	65.83	61.67	63.33	62.50	65.83	60.83
B03	78.12	64.38	77.50	53.75	75.63	54.38	77.50	53.75	76.25	53.75
B04	98.12	91.25	98.75	85.00	98.13	91.88	98.13	91.88	97.50	86.88
B05	60.00	91.25	61.88	84.38	60.63	82.50	61.88	87.50	63.75	77.50
B06	75.00	70.63	75.63	70.00	75.63	74.38	75.63	73.13	74.38	66.25
B07	66.87	60.00	65.00	65.63	66.25	65.63	66.88	62.50	67.50	63.75
B08	90.00	86.88	88.13	86.88	90.00	87.50	90.00	88.13	89.38	87.50
B09	83.75	87.50	85.00	83.13	84.38	86.88	84.38	86.88	86.25	85.00
Average	76.55	75.44	75.95	72.06	76.48	74.35	76.41	74.03	76.48	71.27

TABLE 5.6 Evaluation 5E with Classifier Parameters Obtained from Training Data with Five-Fold CV (neuronal centers spanning −20 to +20 spatial region)

Subject	LDA		SVM		REG		LD4		LD5	
	RQNN	Raw	RQNN	Raw	RQNN	Raw	RQNN	Raw	RQNN	Raw
B01	56.88	50.63	54.38	50.63	51.88	50.63	51.88	50.63	59.38	50.63
B02	60.00	55.63	61.88	55.00	62.50	52.50	60.00	56.25	63.13	55.63
B03	89.38	60.00	90.00	61.88	89.38	65.63	89.38	64.38	91.25	62.50
B04	82.50	70.00	92.50	67.50	82.50	68.13	82.50	72.50	90.00	67.50
B05	69.38	88.75	63.75	85.63	78.13	87.50	62.50	88.13	61.25	86.88
B06	70.63	70.00	77.50	74.38	71.88	75.63	71.88	75.00	73.75	73.75
B07	61.25	65.00	61.25	66.25	61.88	66.25	60.63	66.25	60.00	65.00
B08	94.38	90.63	92.50	88.13	93.75	91.25	92.50	90.63	93.13	90.63
B09	79.38	83.13	79.38	83.75	79.38	83.13	79.38	82.50	78.75	81.88
Average	73.75	70.42	74.79	70.35	74.58	71.18	72.29	71.81	74.51	70.49

5.3.2 Scaling the EEG Prior to Filtering (Reduced Number of Spatial Neurons)

The number of neurons in the RQNN model was reduced to 61, i.e., $N = 60$. Thus, if the input EEG signal varies beyond the $0-2$ region, the spatial neurons can envelop the region beyond this, i.e., from -3 to $+3$[1]. Since the number of neurons along the spatial dimension has been reduced, the filtering process becomes quicker and hence more computationally efficient. To understand this, consider 61 slits which can receive a particular input sample. This can be considered analogous to the double-slit experiment where a beam of light was passed through two slits. Here, 61 neurons act as 61 slits to receive the incoming EEG sample. These neurons cover an input signal varying in the range from -3 to $+3$. The parameter $\hbar$ (reduced Planck's constant) has been set to unity and the other three parameters used to obtain the filtered signal have been set heuristically as $\beta = 2.7$, $m = 0.25$, $\zeta = 15$ and $\beta_d = 1$ for the RQNN with nonlinear modulation of the potential field. In addition, the number of iterative steps that are required for the response of the wave equation to reach a steady state to any particular computational sampling instant of the EEG has been kept to 20. Thus, a particular sample is iterated 20 times before the next sample is presented. The value of the weight and the potential function evolve in this loop. A methodology for appropriately selecting the parameter values is discussed in Section 5.4. However, at this stage, the above parameters are selected heuristically after suitable experimentation with a small EEG dataset based on the work carried out in Chapter 3, and then kept the same for all nine subjects.

Figure 5.5 displays the tracking of the EEG signal in the form of snapshots of the wave packets. The movement of the wave packet along the X-axis is shown at time instants $t = 5.0$ s, $t = 5.2$ s, $t = 5.6$ s and $t = 6.0$ s for a single trial EEG of a representative subject. The Maximum Likelihood Estimate (MLE) from the wave packet gives the filtered EEG shown in Figure 5.4. This figure displays a representative plot of the raw EEG and the RQNN enhanced EEG for a time interval between 5 and 6 s. The effect of filtering can be ascertained through performance enhancement, say by measuring the enhancement in CA and/or kappa values by utilizing the RQNN filtered EEG for further feature extraction and classification process (cf. Figure 5.1).

---

[1]The EEG training dataset has been scaled to lie within the $0-2$ region. However, the EEG during the evaluation stage may vary beyond the $+2$ or below the 0 value. The -3 to $+3$ range of the neuronal lattice is therefore selected to cover the variablity of the EEG data during the unknown evaluation stage.

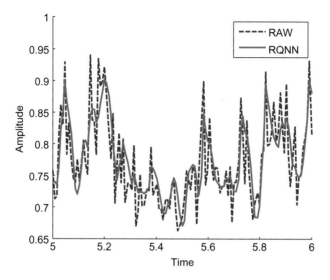

FIGURE 5.4   Representative plot of the RQNN filtered and raw EEG.

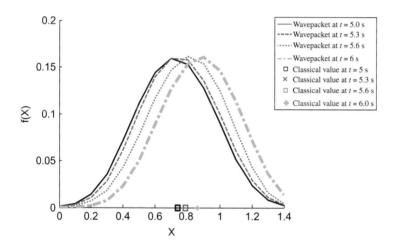

FIGURE 5.5   Snapshots of the wave packets and MLE which generate the representative plot of the RQNN filtered EEG as shown in Figure 5.4.

Tables 5.7 to 5.12 numerically illustrate the maximum CA and kappa values obtained with different classifiers (LDA, SVM, Regression and LD5 (motivated by CSP)) with (cf. Section 2.9.1) and without five-fold CV. The classifier parameters in the five-fold CV are obtained at the most stable time-point where the maximum mean-CA and mean-Kappa coefficients are calculated from the CA and kappa rates obtained on the

TABLE 5.7   CA on 3T Training Dataset

(A) With five-fold cross-validation

Subject	LDA		SVM		REG		LD5	
	RQNN	Raw	RQNN	Raw	RQNN	Raw	RQNN	Raw
B01	77.50	78.13	77.50	79.38	77.50	76.25	77.50	78.13
B02	76.88	59.38	75.63	65.63	74.38	63.75	77.50	58.75
B03	72.50	66.25	70.63	61.88	69.38	57.50	71.88	60.00
B04	99.38	96.25	98.75	95.00	99.38	95.63	98.75	94.38
B05	80.00	80.00	78.13	76.88	81.25	76.25	79.38	78.13
B06	75.00	69.38	77.50	61.25	73.13	70.00	73.75	62.50
B07	89.38	68.13	90.00	66.88	88.13	68.13	89.38	70.00
B08	87.50	83.75	90.00	87.50	87.50	84.38	86.88	85.63
B09	85.63	86.88	84.38	86.25	85.63	86.88	83.13	84.38
Average	82.64	76.46	82.50	75.63	81.81	75.42	82.01	74.65

(B) Without cross-validation

Subject	LDA		SVM		REG		LD5	
	RQNN	Raw	RQNN	Raw	RQNN	Raw	RQNN	Raw
B01	78.75	81.25	79.38	80.00	78.75	81.25	78.75	81.25
B02	78.75	67.50	79.38	69.38	78.75	67.50	79.38	65.63
B03	75.00	68.75	75.00	67.50	75.00	68.13	75.00	66.88
B04	99.38	96.88	98.13	95.63	99.38	96.88	98.75	95.63
B05	84.38	80.63	83.13	77.50	84.38	78.75	83.13	80.00
B06	76.25	73.13	77.50	73.13	76.25	73.13	76.88	70.00
B07	90.63	70.00	90.63	71.25	90.63	70.00	90.63	70.00
B08	90.63	86.25	90.00	87.50	90.63	86.25	90.00	86.25
B09	88.75	87.50	87.50	85.00	88.75	87.50	84.38	87.50
Average	84.72	79.10	84.51	78.54	84.72	78.82	84.10	78.13

five test partitions of the five folds. An average value of the test partitions from these five folds is displayed in the results table for the training dataset. The stable classifier parameters from the five-fold CV are utilized to obtain the evaluation stage performance. The parameters of the classifier without the five-fold CV are obtained from the training set without any CV for setting up the classifier. Figures 5.6 and 5.7 portray

TABLE 5.8   CA on 4E Evaluation Dataset

(A) With classifier parameters obtained from training data with five-fold CV

Subject	LDA		SVM		REG		LD5	
	RQNN	Raw	RQNN	Raw	RQNN	Raw	RQNN	Raw
B01	69.38	61.25	67.50	58.13	68.75	64.38	66.25	63.75
B02	63.33	65.83	63.33	61.67	64.17	61.67	63.33	60.83
B03	76.88	64.38	76.88	53.75	76.88	54.38	76.88	59.38
B04	98.13	91.25	98.75	85.00	98.13	91.88	97.50	86.88
B05	59.38	91.25	63.75	84.38	60.00	82.50	68.13	91.88
B06	75.00	70.63	75.63	70.00	77.50	74.38	74.38	65.63
B07	66.25	60.00	66.25	65.63	67.50	65.63	67.50	63.75
B08	89.38	86.88	88.13	86.88	90.00	87.50	90.63	87.50
B09	83.75	87.50	85.00	83.13	85.63	86.88	85.00	85.00
Average	75.72	75.44	76.13	72.06	76.50	74.35	76.62	73.84

(B) With classifier parameters obtained from training data without CV

Subject	LDA		SVM		REG		LD5	
	RQNN	Raw	RQNN	Raw	RQNN	Raw	RQNN	Raw
B01	68.75	71.25	68.13	65.00	68.75	60.00	67.50	59.38
B02	66.67	61.67	62.50	59.17	66.67	60.83	67.50	60.83
B03	76.88	65.00	76.88	56.25	76.88	56.25	75.63	54.38
B04	98.13	91.88	98.75	85.00	98.13	91.88	97.50	88.13
B05	60.00	92.50	61.25	86.88	60.00	83.75	60.63	79.38
B06	75.63	67.50	75.63	64.38	75.63	65.63	74.38	65.00
B07	66.25	63.13	66.25	63.13	66.25	63.13	66.88	63.75
B08	90.00	86.88	88.75	86.88	90.00	87.50	90.63	87.50
B09	84.38	86.88	85.00	83.75	84.38	86.88	84.38	85.63
Average	76.30	76.30	75.90	72.27	76.30	72.87	76.11	71.55

a graphical representation of the performance of the classifier in terms of classification accuracy values using a bar graph display along with the average performance of all the subjects for each of the classifiers.

It can be concluded from the graphs and the tables that the average performance of all the classifiers is consistent in the training and the evaluation stages. However, the accuracy of the LDA classifier is better

TABLE 5.9   CA on 5E Evaluation Dataset

(A) With classifier parameters obtained from training data with five-fold CV

Subject	LDA		SVM		REG		LD5	
	RQNN	Raw	RQNN	Raw	RQNN	Raw	RQNN	Raw
B01	59.38	50.63	53.13	50.63	59.38	50.63	54.38	50.63
B02	60.00	55.63	60.63	55.00	60.00	52.50	60.00	55.63
B03	88.13	60.00	88.75	61.88	90.00	65.63	90.63	66.25
B04	83.13	70.00	92.50	67.50	83.13	68.13	90.63	68.13
B05	71.88	88.75	63.75	85.63	68.13	87.50	64.38	88.75
B06	70.63	70.00	73.13	74.38	73.13	75.63	74.38	73.13
B07	61.25	65.00	61.25	66.25	60.63	66.25	60.63	65.00
B08	92.50	90.63	92.50	88.13	93.75	91.25	93.13	90.00
B09	79.38	83.13	78.75	83.75	80.00	83.13	81.25	81.25
Average	74.03	70.42	73.82	70.35	74.24	71.18	74.38	70.97

(B) With classifier parameters obtained from training data without CV

Subject	LDA		SVM		REG		LD5	
	RQNN	Raw	RQNN	Raw	RQNN	Raw	RQNN	Raw
B01	58.75	50.63	56.25	50.63	58.75	50.63	59.38	50.63
B02	62.50	53.75	60.00	55.63	62.50	53.75	63.75	55.63
B03	88.75	60.00	88.75	66.25	88.75	66.25	90.63	65.63
B04	83.13	71.88	94.38	65.63	83.13	71.88	90.63	62.50
B05	68.75	89.38	66.88	86.25	68.75	86.25	63.13	86.88
B06	71.88	67.50	73.13	65.63	71.88	65.63	74.38	71.25
B07	60.63	64.38	61.25	65.00	60.63	64.38	60.00	65.00
B08	92.50	90.00	92.50	86.25	92.50	89.38	93.13	88.75
B09	80.00	83.13	80.63	83.13	80.00	83.13	81.25	81.25
Average	74.10	70.07	74.86	69.38	74.10	70.14	75.14	69.72

when using the RQNN model and the raw EEG than when using the other classifiers. Thus, the LDA classifier is chosen for further comparative analysis.

The revised RQNN approach is also investigated on the BCI competition IV 2a dataset [339]. This dataset consists of one training set and one evaluation set for nine subjects with 22 channels and four different MI

TABLE 5.10   Max. of Kappa on Training 3 T Dataset

(A) With five-fold cross-validation

Subject	LDA		SVM		REG		LD5	
	RQNN	Raw	RQNN	Raw	RQNN	Raw	RQNN	Raw
B01	0.54	0.55	0.52	0.58	0.55	0.53	0.54	0.56
B02	0.52	0.19	0.50	0.23	0.47	0.27	0.54	0.18
B03	0.45	0.32	0.41	0.23	0.39	0.15	0.42	0.22
B04	0.99	0.92	0.97	0.90	0.99	0.91	0.97	0.88
B05	0.59	0.60	0.54	0.52	0.63	0.52	0.58	0.56
B06	0.50	0.39	0.51	0.22	0.46	0.40	0.47	0.25
B07	0.78	0.34	0.80	0.34	0.76	0.36	0.79	0.39
B08	0.74	0.67	0.80	0.74	0.75	0.68	0.73	0.71
B09	0.71	0.73	0.68	0.71	0.70	0.74	0.65	0.67
Average	0.65	0.52	0.64	0.50	0.63	0.51	0.63	0.49

(B) Without CV

Subject	LDA		SVM		REG		LD5	
	RQNN	Raw	RQNN	Raw	RQNN	Raw	RQNN	Raw
B01	0.58	0.63	0.59	0.60	0.58	0.63	0.58	0.63
B02	0.58	0.35	0.59	0.39	0.58	0.35	0.59	0.31
B03	0.50	0.38	0.50	0.35	0.50	0.36	0.50	0.34
B04	0.99	0.94	0.96	0.91	0.99	0.94	0.98	0.91
B05	0.69	0.61	0.66	0.55	0.69	0.58	0.66	0.60
B06	0.53	0.46	0.55	0.46	0.53	0.46	0.54	0.40
B07	0.81	0.40	0.81	0.43	0.81	0.40	0.81	0.40
B08	0.81	0.73	0.80	0.75	0.81	0.73	0.80	0.73
B09	0.78	0.75	0.75	0.70	0.78	0.75	0.69	0.75
Average	0.69	0.58	0.69	0.57	0.69	0.58	0.68	0.56

tasks; namely imagining the movement of the left hand (class 1), right hand (class 2), both feet (class 3), and tongue (class 4). The trial began at $t = 0$ s with a fixation cross appearing on the blank screen. In addition, a short acoustic warning tone was presented. After two seconds ($t = 2$ s), a cue in the form of an arrow pointing either to the left, right, down or up (corresponding to one of the four classes left hand, right hand, foot or

TABLE 5.11   Max. of Kappa on 4E Evaluation Dataset

(A) With classifier parameters obtained from training data with five-fold CV

Subject	LDA		SVM		REG		LD5	
	RQNN	Raw	RQNN	Raw	RQNN	Raw	RQNN	Raw
B01	0.39	0.23	0.35	0.16	0.38	0.29	0.33	0.28
B02	0.27	0.32	0.27	0.23	0.28	0.23	0.27	0.22
B03	0.54	0.29	0.54	0.08	0.54	0.09	0.54	0.19
B04	0.96	0.83	0.98	0.70	0.96	0.84	0.95	0.74
B05	0.19	0.83	0.28	0.69	0.20	0.65	0.36	0.84
B06	0.50	0.41	0.51	0.40	0.55	0.49	0.49	0.31
B07	0.33	0.20	0.33	0.31	0.35	0.31	0.35	0.28
B08	0.79	0.74	0.76	0.74	0.80	0.75	0.81	0.75
B09	0.68	0.75	0.70	0.66	0.71	0.74	0.70	0.70
Average	0.51	0.51	0.52	0.44	0.53	0.49	0.53	0.48

(B) With classifier parameters obtained from training data without CV

Subject	LDA		SVM		REG		LD5	
	RQNN	Raw	RQNN	Raw	RQNN	Raw	RQNN	Raw
B01	0.38	0.43	0.36	0.30	0.38	0.20	0.35	0.19
B02	0.33	0.23	0.25	0.18	0.33	0.22	0.35	0.22
B03	0.54	0.30	0.54	0.13	0.54	0.13	0.51	0.09
B04	0.96	0.84	0.98	0.70	0.96	0.84	0.95	0.76
B05	0.20	0.85	0.23	0.74	0.20	0.68	0.21	0.59
B06	0.51	0.35	0.51	0.29	0.51	0.31	0.49	0.30
B07	0.33	0.26	0.33	0.26	0.33	0.26	0.34	0.28
B08	0.80	0.74	0.78	0.74	0.80	0.75	0.81	0.75
B09	0.69	0.74	0.70	0.68	0.69	0.74	0.69	0.71
Average	0.53	0.53	0.52	0.45	0.53	0.46	0.52	0.43

tongue) appeared and stayed on the screen for 1.25 s. This prompted the subjects to perform the desired motor imagery task. However, the RQNN approach was carried out, as before, using only two channels, namely C3 and C4, and only for a two-class classification (left hand vs. right hand). Therefore, the data were separated into two classes, EEG with left hand and right hand mental imagination task. The parameters

TABLE 5.12 Max. of Kappa on 5E Evaluation Dataset

(A) With classifier parameters obtained from training data with five-fold CV

Subject	LDA		SVM		REG		LD5	
	RQNN	Raw	RQNN	Raw	RQNN	Raw	RQNN	Raw
B01	0.19	0.01	0.06	0.01	0.19	0.01	0.09	0.01
B02	0.20	0.11	0.21	0.10	0.20	0.05	0.20	0.11
B03	0.76	0.20	0.78	0.24	0.80	0.31	0.81	0.33
B04	0.66	0.40	0.85	0.35	0.66	0.36	0.81	0.36
B05	0.44	0.78	0.28	0.71	0.36	0.75	0.29	0.78
B06	0.41	0.40	0.46	0.49	0.46	0.51	0.49	0.46
B07	0.23	0.30	0.23	0.33	0.21	0.33	0.21	0.30
B08	0.85	0.81	0.85	0.76	0.88	0.83	0.86	0.80
B09	0.59	0.66	0.58	0.68	0.60	0.66	0.63	0.63
Average	0.48	0.41	0.48	0.41	0.48	0.42	0.49	0.42

(B) With classifier parameters obtained from training data without CV

Subject	LDA		SVM		REG		LD5	
	RQNN	Raw	RQNN	Raw	RQNN	Raw	RQNN	Raw
B01	0.18	0.01	0.13	0.01	0.18	0.01	0.19	0.01
B02	0.25	0.08	0.20	0.11	0.25	0.08	0.28	0.11
B03	0.78	0.20	0.78	0.33	0.78	0.33	0.81	0.31
B04	0.66	0.44	0.89	0.31	0.66	0.44	0.81	0.25
B05	0.38	0.79	0.34	0.73	0.38	0.73	0.26	0.74
B06	0.44	0.35	0.46	0.31	0.44	0.31	0.49	0.43
B07	0.21	0.29	0.23	0.30	0.21	0.29	0.20	0.30
B08	0.85	0.80	0.85	0.73	0.85	0.79	0.86	0.78
B09	0.60	0.66	0.61	0.66	0.60	0.66	0.63	0.63
Average	0.48	0.40	0.50	0.39	0.48	0.40	0.50	0.39

chosen for the RQNN model were $\beta = 3$, $m = 0.5$ and $\zeta = 15$. The number of neurons along the X-axis was fixed at $N = 100$ while each sample was recurrently iterated six times in the SWE loop. This is necessary, as discussed earlier, to stabilize the SWE integration loop. The de-learning parameter β_d, as previously, was fixed at 1. All these parameters were selected after suitable trial and error experimentation with a small set of

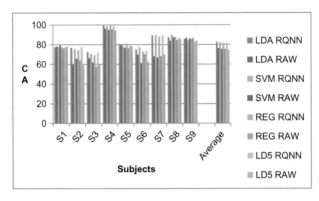

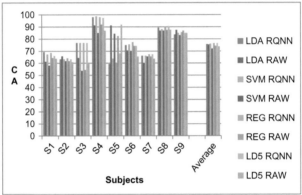

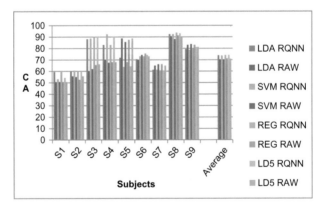

FIGURE 5.6   Training and evaluation with five-fold cross-validation.

EEG data. Table 5.13 displays the CA and kappa values for the RQNN filtered and raw EEG signals obtained using the two channels. The average CA and kappa show an overall improvement of 3–4% and 0.07 respectively through use of the RQNN model. However, the parameters

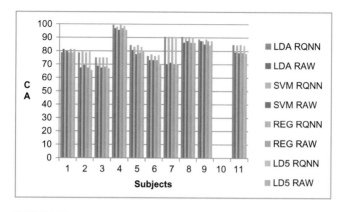

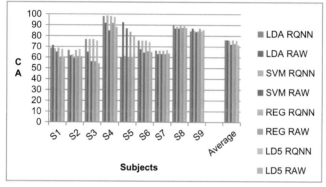

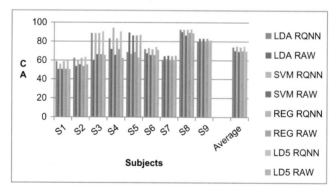

FIGURE 5.7   Training and evaluation without cross-validation.

of the RQNN model are the same for all subjects. In practice, a parameter suitable for one subject may not give the desired result for another. Therefore, it is important to tune the RQNN parameters subject-specifically. This is discussed and investigated in the next section.

TABLE 5.13 RQNN Performance on BCI Competition IV 2A Dataset

Subject	Accuracy (Training)		Kappa (Training)		Accuracy (Evaluation)		Kappa (Evaluation)	
	RQNN	Raw	RQNN	Raw	RQNN	Raw	RQNN	Raw
A01	67.36	68.05	0.35	0.36	63.19	61.11	0.26	0.22
A02	69.44	61.80	0.39	0.24	55.55	52.77	0.11	0.06
A03	75.00	81.25	0.50	0.63	70.13	76.38	0.40	0.53
A04	65.97	65.97	0.32	0.32	59.72	59.72	0.19	0.19
A05	79.86	66.66	0.60	0.33	70.13	50.00	0.40	0.01
A06	70.13	69.44	0.40	0.39	65.27	68.05	0.31	0.36
A07	82.63	71.52	0.65	0.43	82.63	61.80	0.65	0.24
A08	73.61	65.97	0.47	0.32	56.25	57.63	0.13	0.15
A09	79.16	82.63	0.58	0.65	82.63	86.11	0.65	0.72
Average	73.68	70.37	0.47	0.41	67.28	63.73	0.35	0.28

5.4 TOWARDS SUBJECT-SPECIFIC RQNN PARAMETERS

This section discusses one of the possible ways of tuning the RQNN parameters to suit each individual subject. Until now, RQNN parameters have been obtained heuristically and then kept the same for all the subjects. This section discusses an approach to obtaining subject-dependent RQNN parameters which could lead to a better cross-session stable performance. One such approach is to utilize optimization techniques such as PSO or GA discussed in Section 2.9 and applied to EEG filtering in reference [213]. However, as discussed earlier, there are several parameters to tune and hence applying any optimization technique without knowledge of the landscape of the multi-dimensional search space would be difficult. There is a need to determine where the optimum lies by evaluating performance with every possible parameter combination. Therefore, this section discusses a relatively manual approach for parameter tuning by using a two-step inner-outer five-fold CV technique [327]. The sole purpose of utilizing such a manual approach is to obtain a parameter set which results in stable performance for the EEG signal classification across different sessions.

5.4.1 Two-Step Inner-Outer Five-Fold Cross-Validation for RQNN Parameter Selection

5.4.1.1 The Method

RQNN parameter selection is performed by using an inner-outer five-fold CV in a two-step process (see flowchart in Figure 5.8). The first step consists of finding a combination of Hjorth and bandpower features using the standard frequency bands of 8−12 Hz and 16−24 Hz (for bandpower feature extraction [FE]). We aim to determine four parameters in the RQNN model using the inner-outer five-fold CV, while the remaining parameters remain fixed. The fixed set of parameters and their explanation are given in Table 5.14. They are the number of neurons and their spacing within the spatial axis, as well as the parameters relating to setting up iterations within the SWE loop. These parameters may not be globally optimal; however, their values are kept in accordance with the range of the varying input EEG signal. The variable parameters and their range during the two-step inner-outer five-fold CV are detailed in Table 5.15. These parameters relate to the learning rules applied for the filtering process of the RQNN, i.e., the learning and the de-learning rate update of the synaptic weight vector while the scaling factor and the mass of the quantum object are associated with the excitation of the potential field and the SWE that ultimately affects the movement of the wave packet. These parameters are varied within a suitable range in the first step of the parameter selection approach. In this step, the complete EEG data are divided into five outer folds. Of these, the RQNN and the FE are applied with each of the parameter combinations on four folds. Next, five inner folds are created from the four outer folds and an inner five-fold CV is performed at the classification stage. Thus, the data are partitioned into a training set (80% of the data) and a test set (20% of the data) which is used to calculate the CA rates. The mean-CA i.e., $\frac{1}{5}\sum_{i=1}^{5} CA_i$ and mean-Kappa coefficient i.e., $\frac{1}{5}\sum_{i=1}^{5} Kappa_i$ are calculated from the CA and kappa rates obtained on the five test partitions. This procedure is performed on all the outer folds. This first step therefore gives one best RQNN parameter combination for each outer fold. The second step begins with a set of these five best RQNN parameters from each of the outer folds. Next is to find the best frequency band in consultation with the five sets of the RQNN parameters. Here, 12 different combinations of frequency bands varying within the 8−26 Hz frequency region are utilized as a search space. All the combinations of the five RQNN parameters and the 12 frequency bands are used to obtain the enhanced EEG and features on the complete training dataset of the MI EEG. A normal five-fold CV is performed to obtain the time-point with the highest mean-Kappa coefficient

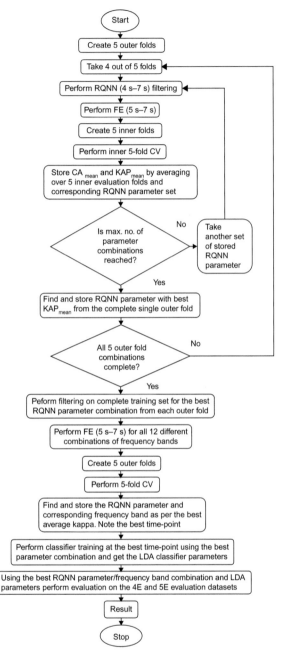

FIGURE 5.8   Flowchart for parameter tuning (RQNN/frequency band). *Reproduced from [327].*

TABLE 5.14   Fixed RQNN Parameters Prior to Initializing the Variable Parameter Search

Parameter	Explanation	Value
N	It is the number of neuronal centers along the spatial axis of the RQNN model. This can be considered as if some quantum object (i.e., each neuron in the form of a window) in the neuronal lattice is looking at the spatial neural network and accepting the input signal sample	61
MAXIT	It is the number of outer loop iterative steps that are required for the response of the SWE to reach a steady state for a given computational sampling instant	10
iLoop	It is the number of inner loop iterative steps that are required for the response of the SWE to reach a steady state for a given computational sampling instant	5
δt	The SWE equation is converted to the finite difference form with time steps δt	0.001
δx	The spatial axis is divided into N mesh points (here 60) so that x is represented as $x_j = j\delta x$ where j varies from $-\frac{N}{2}$ to $+\frac{N}{2}$. As δx was fixed as 0.1, it covers the spatial axis range from -3 to $+3$	0.1

Reproduced with permission from [327].

TABLE 5.15   Variable Parameters to be Tuned Within the Search Space

Parameter	Explanation	Search Space
β	This is a learning parameter and it is necessary to update the synaptic weight vector $W(x,t)$	0.5–9
m	This is the mass of the quantum object and it is associated with self-excitation of the SWE	0.05–1
ζ	This is the scaling factor to actuate the spatial potential field $V(x)$ and it causes input excitation (since it appears as a multiplicand in the SWE). The value of ζ can be negative or positive	15–9500
β_d	This is the de-learning parameter, and it is used to forget previous information. De-learning is used to prevent unbounded increase of the values of the synaptic weights $W(x,t)$	1–8

Reproduced with permission from [327].

value (average of the five test partitions). The RQNN parameters and frequency band combination are chosen to correspond with the highest mean-Kappa coefficient value. The classifier is trained at the best time-point on the complete training dataset with this RQNN parameters and frequency band combination. This completes the second stage of the parameter selection process, and it is repeated for all subjects and applied to the evaluation set.

5.4.1.2 Results and Analysis

Table 5.16 displays the best RQNN and frequency band parameters obtained from the two-step inner-outer five-fold CV for all the nine subjects of the BCI competition IV 2b dataset. The predetermined parameters shown in Table 5.14 are kept fixed for all the nine subjects.

The performance in terms of maximum of CA and kappa obtained with the RQNN is compared with the unfiltered EEG and the SG filtered EEG. The same is displayed in Table 5.17 and Table 5.18 for the BCI competition IV 2b dataset. The frequency bands for the raw EEG and the SG filtered approach were optimally obtained in the same way as that with the second stage of the RQNN parameter selection process. Suitable parameters for the SG technique were obtained heuristically in the present work. The average improvement with the RQNN technique across all the nine subjects is more than 4% in CA and 0.08 in kappa values when compared with the raw or the SG filtered approach. These results also show a clear improvement of more than 9% in average CA and more than 0.1 in average kappa value when compared with the BCI design with power spectral density (PSD) features extracted from raw EEG investigated in reference [343] on the same dataset. In addition, the RQNN is also the clear winner with improvements of about 4% in the average CA and more than 0.07 in average kappa when compared with

TABLE 5.16   Subject-Specific RQNN Parameters from Inner-Outer Five-Fold CV for the BCI Competition IV 2B Subjects

Subject	RQNN parameters				Frequency bands			
	β	m	ζ	β_d	lf_1	hf_1	lf_2	hf_2
B01	6	0.25	1815	1	8	10	16	20
B02	2	0.7	315	8	11	13	16	24
B03	5	0.2	415	7	8	10	14	16
B04	3	0.8	215	7	8	12	16	24
B05	8	0.7	115	3	8	12	14	16
B06	2.5	0.65	215	6	11	13	14	16
B07	5	0.75	115	8	8	12	14	16
B08	5.5	0.7	15	3	11	13	14	16
B09	2.5	0.6	500	8	8	12	16	24

Reproduced with permission from [327].

TABLE 5.17 CA on BCI Competition IV 2B Dataset (subject-specific parameters obtained through inner-outer five-fold CV)

Subject	Training (03T)			Evaluation (04E)			Evaluation (05E)		
	RQNN	SG	RAW	RQNN	SG	RAW	RQNN	SG	RAW
B01	88.13	83.13	85.63	79.38	65.63	76.88	75.00	64.38	73.75
B02	81.88	80.63	81.88	62.50	63.33	60.83	63.75	63.75	61.88
B03	93.13	79.38	80.63	77.50	72.50	82.50	94.38	80.63	91.88
B04	100.00	96.88	99.38	99.38	96.88	98.75	96.25	90.63	86.25
B05	92.50	80.63	88.75	88.75	76.25	80.00	93.75	76.25	80.00
B06	85.63	78.13	77.50	81.25	75.63	76.88	88.75	75.63	80.00
B07	92.50	90.63	88.75	73.75	75.63	70.63	71.25	68.75	66.25
B08	97.50	88.13	90.00	86.25	85.63	86.25	96.88	91.25	96.88
B09	87.50	83.13	87.50	86.88	74.38	79.38	88.13	75.00	78.75
Avg.	90.97	84.51	86.67	81.74	76.20	79.12	85.35	76.25	79.51

Reproduced with permission from [327].

the performance of the BCI design with Bispectrum features extracted from raw EEG investigated in reference [343] on the same dataset. It can therefore be firmly concluded from these results that RQNN improves the average performance of the BCI system for almost all the subjects during both the training and the evaluation stages when compared with the unfiltered EEG, SG filtered EEG and even PSD and Bispectrum feature-based approaches. In fact, the enhancement in performance is significantly better than those obtained by the winner of the BCI competition discussed in reference [341].

To determine whether these performance improvements are statistically significant, a two-way analysis of variance (ANOVA2) test is performed on the results of the training and the evaluation stages for the RQNN filtered and the raw EEG approach. p-Values of 0.0166 (3T), 0.0920 (4E), 0.0065 (5E), 0.0217 (combined 4E and 5E) were obtained for the accuracy, and p-values of 0.0151 (3T), 0.0921 (4E), 0.0065 (5E) and 0.0216 (combined 4E and 5E) were obtained for the kappa. As all but one of the p-values are much less than 0.05, the null hypobook can be rejected and the alternative hypobook that the results from the RQNN filtered EEG are likely to be different from the result of the raw EEG, can be accepted, i.e., the improvements due to the subject-specific RQNN model-based technique are statistically significant.

The two-step inner-outer five-fold CV was also applied to the BCI competition IV 2a dataset. The subject-specific RQNN parameters

TABLE 5.18 Max. Kappa on BCI Competition IV 2B Dataset (subject-specific parameters obtained through inner-outer five-fold CV)

Subject	Training (03 T)			Evaluation (04E)			Evaluation (05E)		
	RQNN	SG	RAW	RQNN	SG	RAW	RQNN	SG	RAW
B01	0.7570	0.6615	0.7079	0.5875	0.3125	0.5375	0.5000	0.2875	0.4750
B02	0.6338	0.6084	0.6368	0.2500	0.2667	0.2167	0.2750	0.2750	0.2375
B03	0.8603	0.5853	0.6087	0.5500	0.4500	0.6500	0.8875	0.6125	0.8375
B04	1.0000	0.9371	0.9875	0.9875	0.9375	0.9750	0.9250	0.8125	0.7250
B05	0.8494	0.6113	0.7732	0.7750	0.5250	0.6000	0.8750	0.5250	0.6000
B06	0.7091	0.5559	0.5519	0.6250	0.5125	0.5375	0.7750	0.5125	0.6000
B07	0.8495	0.8085	0.7751	0.4750	0.5125	0.4125	0.4250	0.3750	0.3250
B08	0.9498	0.7497	0.7982	0.7250	0.7125	0.7250	0.9375	0.8250	0.9375
B09	0.7524	0.6545	0.7404	0.7375	0.4875	0.5875	0.7625	0.5000	0.5750
Avg.	0.8179	0.6858	0.7311	0.6347	0.5241	0.5824	0.7069	0.5250	0.5903

TABLE 5.19   Subject-Specific Parameters for BCI Competition IV 2A Subjects

Subject	RQNN Parameters				Frequency Bands			
	β	m	ζ	β_d	lf_1	hf_1	lf_2	hf_2
A01	3.5	0.8	715	4	8	10	16	20
A02	6.5	0.75	715	2	8	12	20	24
A03	2.5	0.2	415	2	11	13	14	16
A04	5.5	0.7	715	7	11	13	16	24
A05	5	0.5	415	4	8	10	16	24
A06	6	0.7	615	3	8	10	14	16
A07	5	0.1	815	2	8	12	16	24
A08	5	0.55	115	5	8	10	16	24
A09	3	0.65	915	3	8	10	16	24

obtained for this dataset are displayed in Table 5.19. It should be mentioned here that only the variable set of parameters has been selected through the two-step procedure, while the fixed set of parameters has been kept the same as in Table 5.14. The performance in terms of CA and maximum of kappa obtained with the RQNN is compared with the unfiltered EEG in Table 5.20 and the enhancement in performance is evident with the RQNN filtered EEG [327].

5.4.1.3 Concluding Remarks

In conclusion, it can be stated that RQNN can be used as a signal filtering mechanism to enhance the overall performance of the BCI system. The parameters of the model can be selected appropriately through computational algorithms, or a manual approach such as the one performed in this section in the form of inner-outer five-fold CV. However, the parameter selection/tuning should be performed after an appropriate understanding of the contribution of each parameter has been gained, as discussed in Chapter 3.

5.5 DISCUSSION

There are many parameters within the RQNN model that should be selected appropriately to obtain the desired response of the SWE. The complex task of selecting the best combination of the RQNN parameters for EEG filtering is implemented in Section 5.4 using a two-step, inner-outer, five-fold cross-validation. Other optimization techniques such as

TABLE 5.20   RQNN Performance on BCI Competition IV 2A Dataset
(Subject-Specific Parameters)

Subject	Accuracy (Training)		Kappa (Training)		Accuracy (Evaluation)		Kappa (Evaluation)	
	RQNN	Raw	RQNN	Raw	RQNN	Raw	RQNN	Raw
A01	66.60	66.06	0.33	0.32	61.11	61.11	0.22	0.22
A02	67.24	69.48	0.36	0.40	61.11	57.64	0.22	0.15
A03	77.78	75.02	0.55	0.49	79.17	77.78	0.58	0.56
A04	65.32	63.89	0.30	0.28	60.42	56.25	0.21	0.13
A05	70.79	68.03	0.41	0.37	71.53	66.67	0.43	0.33
A06	66.01	62.54	0.32	0.25	61.11	61.11	0.22	0.22
A07	68.67	68.13	0.37	0.37	58.33	56.25	0.17	0.13
A08	67.41	65.86	0.35	0.31	67.36	66.67	0.35	0.33
A09	78.40	73.55	0.56	0.46	79.17	75.00	0.58	0.50
Average	69.80	68.06	0.40	0.36	66.59	64.27	0.33	0.29

Reproduced with permission from [327].

PSO or GA for parameter selection/tuning can also be investigated (cf. Section 2.9 and Section 3.6). However, incorporating such a global optimization methodology for tuning the RQNN parameters for EEG signal filtering will require a huge, multi-dimensional, search space spanning 13 variables, and will be a challenging task. To clearly illustrate the extent of the problem, the two-step inner-outer five-fold cross-validation parameter optimization for each subject implemented on an 84-core set-up took approximately 24 hours of computational time. Therefore, although this interesting but computationally expensive parameter selection task has the potential to enhance the information from the otherwise noisy EEG signals, it will be a demanding effort.

The revised RQNN model is stimulated directly using the raw scaled input signal so that the number of neurons enveloping the noisy input signal is reduced (cf. Section 5.3). This also lowers the computational load on the movement of the wave packet that tracks the noisy input signal by reducing the number of iterations applied to each computational sampling instant. The method can therefore be implemented practically in real-time for EEG signal enhancement (cf. Chapter 6 for further details).

These points outline the potential advantages of the RQNN framework for filtering complex signals such as EEG, but further analysis is required to verify this potential.

5.6 CONCLUSION

This chapter has discussed the application procedures and details that are necessary to implement RQNN models for enhancing the information from noisy EEG datasets. As mentioned at the beginning of this chapter, the development of the RQNN model has followed an evolutionary design cycle. The basics of RQNN models were discussed in Chapter 3. The model was initially implemented without any scaling and with a large number of neuronal centers. This meant that a large spatial area had to be investigated with the wave packet and resulted in a time-consuming wave packet shift. This large spatial area of movement of the wave packet was subsequently reduced and so was the number of neuronal centers spanning the spatial axis. This process of developing the RQNN model has formed the basis for further work in the form of practical implementation of these models for online EEG signal classification in a real-time scenario with lower computational requirements. The parameter selection approach in the form of two-step, inner-outer, five-fold CV has successfully resulted in enhanced BCI system performance, adapted to suit each individual subject. The results have shown notable enhancement in the performance of the BCI system for the two BCI competition datasets even across multiple sessions. This is a big step towards obtaining a subject-specific RQNN model that suits individual subjects, and the enhancements are also statistically significant. The next step is an advanced form of parameter tuning approach using PSO, GA or any computational algorithm for global optimization. The next chapter discusses the practical implementation of the RQNN filtering technique in the BCI laboratory at the Intelligent Systems Research Centre (ISRC), University of Ulster.

6

Graphical User Interface (GUI) and Robot Operation

6.1 INTRODUCTION

This chapter details the practical realization of a complete brain−computer interface (BCI) system with the help of the foundation work discussed in the earlier chapters. The real-time implementation was carried out in the BCI laboratory at the Intelligent Systems Research Center (ISRC), University of Ulster (cf. video in [344]). The work begins with the Recurrent Quantum Neural Network (RQNN) technique as a pre-processing model to filter the unknown amount and type of noise embedded within the electroencephalogram (EEG) signals collected from different subjects. The Hjorth and bandpower features extracted from the RQNN enhanced EEG signal are then fed to the Linear Discriminant Classifier (LDA). The classifier output is post-processed using the multiple threshold concepts discussed earlier, and sent to the brain−robot interface or the graphical user interface (GUI). The single valued outcome from the GUI is interpreted as a final command for maneuvering the mobile robot in any specified direction, such as Forward, Right, etc. in the supervised mode and as a destination-specific command in the form of 'Room 1', 'Cupboard', etc. in the autonomous mode. For performance evaluation, the RQNN model and the interface are evaluated independently of one another. Thus, during the evaluation of the interface, the RQNN model is not a part of the BCI system and vice versa. However, the complete BCI system along with the RQNN model and the adaptive interface are investigated for maneuvering the pioneer mobile robot in real time.

The aim of this chapter is to confirm the enhanced performance of the classifier in terms of classification accuracy (CA) and kappa values when using the RQNN-filtered EEG signal over the raw EEG signal that was acquired from three subjects at the ISRC. In addition, the aim of

151

this chapter is also to compare the theoretical and practical performance of the proposed user-centric iAUI for maneuvering the mobile robot in an online real-time scenario. The performance of the iAUI is compared with the other state-of-the-art interface designs. The chapter begins with a discussion of the EEG acquisition process carried out in the BCI laboratory, in Section 6.2. Section 6.3 details the enhancement that is obtained in BCI performance by filtering the acquired EEG using the RQNN model. Section 6.4 discusses the advantages in terms of cost reductions when the proposed interface designs are used to enhance the overall communication bandwidth of the BCI. This section also compares these designs with the other state-of-the-art interface designs. Section 6.5 discusses the real-time implementation of the adaptive interface design for maneuvering the mobile robot in a simulated environment. Lastly, Section 6.6 discusses real-time maneuvering of the pioneer mobile robot with RQNN filtering of the EEG signals, and while the interface is updated by the sonar sensors mounted on the robot in the physical arena.

6.2 THE EEG ACQUISITION PROCESS

The RQNN model has been investigated by carrying out experiments on three subjects in the BCI laboratory at the ISRC. The EEG signal from two channels (C3 and C4) was recorded in bipolar mode at a sampling frequency of 256 samples/s by utilizing g.tec gUSBamp/gMOBIlab$^+$ equipment. It should be mentioned here that dry electrodes[1] [345] were used for these experiments. The data are collected from the subject in runs consisting of 40 trials, with each trial lasting 7 s/5 s, according to the trial diagram shown in Figure 6.1. The subject is asked to perform mental imagination indicative of either the right hand or the left hand motor imagery (MI), as indicated by the direction in which an arrow points. These directions are randomly generated during each trial by the well-known arrow paradigm.

The EEG signal is enhanced using the RQNN model to filter the unknown noise embedded within the EEG. As discussed in the previous chapters, this is the strength of the RQNN model; it performs the

---

[1]The application of EEG is impeded by the need to prepare the electrodes with conductive gel. This is necessary to lower the impedance between the electrodes and the scalp, but is cumbersome to use. Dry electrodes have the potential to provide a solution to this problem, and reduce the experimental preparation time [382]. However, dry electrodes are still the object of active international research as they still suffer from increased artifact sensitivity and high contact impedance value [383].

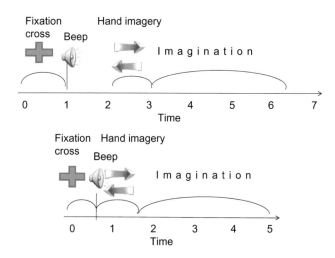

FIGURE 6.1   Training paradigm and timing for 7 s and 5 s trial times.

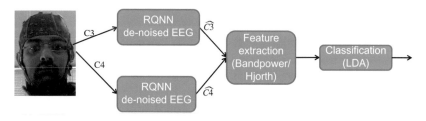

FIGURE 6.2   Framework of the EEG signal estimation using RQNN model.

filtering process without any knowledge of the type or amount of noise that is embedded within the signal. During the process of filtering the signal, the single channel time series data are fed to the RQNN model one sample at a time. Since the work presented here consists of two channels (C3 and C4) of EEG data, each channel data is fed through different RQNNs to obtain the filtered signal. Therefore, the complete setup looks like the one shown in Figure 6.2. Here, the EEG data from the C3 and C4 channels are fed to the two RQNNs, and an estimate of the signal is obtained for the samples from both these channels. After this step, suitable features (here Hjorth [53] and band power (BP) features) are extracted from these signals. These features are then fed as inputs to train the offline classifier, which in the present work is the LDA [221] classifier. Once the offline analysis is complete and the classifiers are trained, the parameters/weight vector of the classifier are stored for later use with the classifiers (within the simulink model shown in Figure 6.3). They will be used to identify the unlabeled EEG

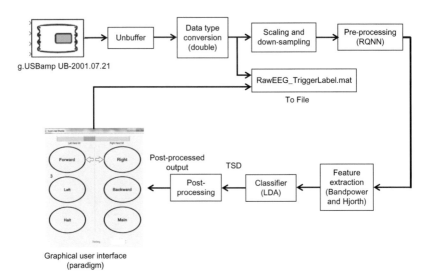

Graphical user interface
(paradigm)

FIGURE 6.3   Complete BCI block diagram within the simulink framework in MATLAB.

signals during online analysis. It should be clarified here that, in order to capture the time series property of the continuous EEG signal, the weight updating process of the RQNN model is continuous (to filter the raw EEG signal) during both the offline as well as the online stages. However, the classifier parameters are tuned offline in order to have fixed weight parameters for access to the online classifier. Therefore, the framework shown in Figure 6.2 will be the same in the classifier training process during the offline stage, or in the evaluation process during the online stage when the classifier parameters have been fixed. The online classification of the EEG signals is carried out continuously. However, the BCI user should perform MI by associating the mental task with an appropriate command icon indicated on the GUI. The GUI ultimately sends the command to the device being controlled. The next section details the evaluation of the performance of the BCI in terms of CA and kappa values using the RQNN model, without considering the GUI stage.

6.3 RQNN-BASED EEG SIGNAL ENHANCEMENT

To incorporate the approaches discussed, the EEG data are first collected and used to train the LDA classifier offline. The parameters obtained from this offline training are then used with the online LDA classifier, an example of which is shown in the simulink model (cf. Figure 6.3). The task begins with the acquisition of EEG signals using the dry electrodes from the gUSBamp, which is displayed as the first component in the model. The raw EEG signals are down-sampled and scaled appropriately, so that they are within the range of the spatial neurons of the subject-specific RQNN model (discussed in Section 5.4). The RQNN model is shown in the pre-processing block in the diagram. The filtered EEG signal (i.e., the output of the RQNN model) is utilized to extract the Hjorth and bandpower features (discussed in Section 2.6) as shown within the feature extraction (FE) block. These output features of the FE block are subsequently utilized in the online LDA classifier (discussed in Section 2.7) to determine the appropriate TSD (time-varying signed distance)/classifier output. This TSD signal is post-processed (discussed in Section 2.8) and sent to the GUI (cf. Chapter 4) using the MATLAB S-function. However, for RQNN model evaluation purposes, this section only evaluates the performance of the RQNN model as a preliminary step without going into the details of the GUI. Therefore, the user is required to perform the MI in accordance with the synchronous arrow paradigm (discussed in Section 2.3.1).

The parameters obtained for the three subjects using the two-step inner-outer five-fold CV parameter selection approach (cf. Section 5.4) are detailed in Table 6.1. The parameter β is the learning parameter, which is necessary to update the synaptic weight vector W, while the parameter β_d is the de-learning parameter in the RQNN model. Similarly, the parameter m is the mass of the quantum object and ζ is a scaling factor to actuate the spatial potential field $V(x)$ in the RQNN model. These four parameters are obtained in the first step of the two-step parameter selection approach. The other four parameters are the frequency bands of the FE component and they are selected in the second step. The parameters lf_1, hf_1, lf_2 and hf_2 are the lower and upper frequency limits of the Mu and Beta bands, respectively. The BCI performance in terms of CA and kappa values for the three subjects is shown in Table 6.2. All three subjects were male, in the 27−35 age group, and had previous experience with the BCI system. The training and the evaluation datasets for the subject V01 were obtained from four runs of the first recording and two runs with a gap of 1 week using the usual 7 s trial time (cf. Figure 6.1) respectively. However, the training and the evaluation datasets for the subject J02 were obtained from two runs of the first recording and two runs with a gap of 2 weeks respectively but using a 5 s trial time (cf. Figure 6.1). Subject J02 was experienced and preferred a shorter trial time to acquaint himself with operating the device by using the proposed interface. The training and the evaluation dataset for subject G03 were obtained from two runs of the first recording and two runs with a gap of 3 days respectively using the usual 7 s trial time. The results in terms of maximum CA and kappa

TABLE 6.1 RQNN Parameters Using Inner-Outer Five-Fold CV for Subjects in the BCI Laboratory

Subject	RQNN parameters				Frequency bands			
	β	m	ζ	β_d	lf_1	hf_1	lf_2	hf_2
V01	4.0	0.25	915	4	8	12	16	24
J02	8.0	0.80	115	6	8	12	20	24
G03	3.5	0.80	515	5	8	10	16	20

TABLE 6.2 RQNN Performance for the Subjects in the BCI Laboratory

Subject	Accuracy (Training)		Kappa		Accuracy (Evaluation)		Kappa	
	RQNN	Raw	RQNN	Raw	RQNN	Raw	RQNN	Raw
V01	68.75	69.37	0.37	0.38	78.75	76.25	0.58	0.53
J02	91.25	90.00	0.82	0.79	87.50	83.75	0.75	0.67
G03	75.00	73.75	0.52	0.49	75.00	72.50	0.50	0.45
Average	78.33	77.70	0.57	0.55	80.41	77.50	0.61	0.55

values show enhancement with respect to the raw EEG signals. This shows that the RQNN efficiently filters the EEG signals, which subsequently enhances the classification performance.

6.4 AUTONOMOUS AND SUPERVISED GUI OPERATION

The post-processed TSD from the BCI user is the final deciding factor in issuing a command to the GUI. However, the online classification of the user EEG signals is in synchronization with the GUI designed for the wheelchair/robot control. The output from the signal processing i.e., the post-processed TSD helps select one of the several possible choices in the GUI (if the value of the TSD is above the positive threshold or below the negative threshold) or the no-control (NC) state (if the value of the TSD is between the positive and the negative thresholds) (cf. Figure 2.12). The GUI, as discussed earlier, is utilized to enhance the otherwise low communication bandwidth of the two classes. The proposed interface for controlling the mobile robot in an unstructured environment is analyzed next by considering 100% accuracy from the signal processing block of the BCI.

6.4.1 Maneuvering the Mobile Robot Under a 100% BCI Accuracy Assumption

The operation of the interface can be explained in detail by considering the diagram of a typical robotic arena within player-stage [335] simulation shown in Figure 6.4. Here, three different locations are identified as entrances to 'Room 1', 'Room 2' and a 'cupboard.' The robot is located in the bottom left corner of the robotic arena within a player-stage environment. The player-stage provides a network interface to a variety of robot and sensor hardware. It simulates a population of mobile robots moving in and sensing a two-dimensional bitmapped environment [346]. Here, a simulated robot within the player-stage environment is chosen rather than an actual robot[2], because the player-stage environment simulates the physical dynamics of the pioneer robot as well as the environment, and thus facilitates easy transition of the results to the real world. Programs are often first developed and debugged in a simulated arena, and then moved to a real-world environment for investigation.

An orange-colored marker identifies all three locations on the simulated robotic arena, i.e., 'Room 1', 'Room 2' and 'Cupboard.' This marker does not have any effect on the movement of the robot, nor is it used for identification of the particular location for the robot. The marker is only there to help the BCI user. The user is required to maneuver the robot to each of the three locations from the origin by using the iAUI, the non-adaptive interface and the autonomous interface. The task of evaluating the performance of the interface is done by measuring the time taken and the number of commands required to reach each of these target locations from the original starting position of the robot (cf. Figure 6.4). For evaluation purposes in this section, an ideal TSD with 100% CA like the one shown in Figure 6.5 is considered. The purpose of using an ideal TSD is to evaluate the interface for a device control task without taking into consideration the accuracy limitations of the signal processing block within the BCI system. The main aim of this part of the work is to thus evaluate the capability of the interface and not of the accuracy of the BCI user, as the latter depends on several factors including user mood swings as well as signal processing. Therefore, the parameters 'mission time' (the total time required for the completion of a task) and 'concentration time' (the time-period required by the user to focus and drive the mobile robot

---

[2]Over 50 different research laboratories and institutions all around the world are currently involved in the active development of the player-stage and even more are using it [335]. Stage devices present a standard Player interface so few or no changes are required to move between simulation and hardware.

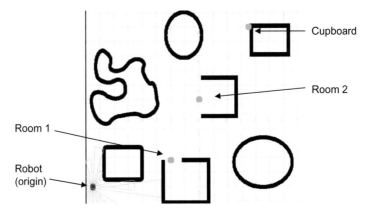

FIGURE 6.4   Screenshot of the robot arena (simulated in player-stage environment).

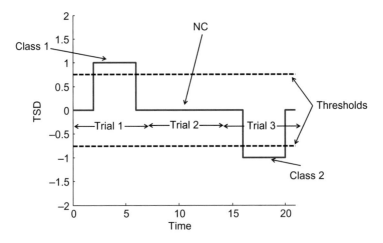

FIGURE 6.5   Ideal TSD with 100% CA assumption.

during the maneuvering phase) are used for evaluation. Both these terms are discussed by Rebsamen et al. [300] and are also referred to in Section 2.12.

In the work presented in this book, the concentration time is measured as the length of time required by the BCI user performing MI to issue the commands that are required to perform any given task. The concentration time excludes the time during which the user does not issue any command in a scan cycle. The interface is evaluated once by utilizing an adaptive feature in the supervised control interface, once without the adaptive feature in the supervised control interface and lastly with the autonomous control interface (no adaptation) to reach the three locations. Before detailing the evaluation process, each of the interfaces is revisited briefly.

Non-adaptive interface: This is the most basic form of the supervised interface in which the different tasks/choices offered to the BCI user remain in fixed positions, i.e., the interface does not adapt (cf. Figure 4.3).

Adaptive interface: This is an adaptive version of the supervised interface, i.e., the iAUI. The different tasks/choices available to the BCI user are re-positioned dynamically within the interface in real-time in response to the changing environment, and thereby assist the user and allow faster communication (cf. Figure 4.7).

Autonomous interface: This is an unsupervised form of the interface which displays the available tasks/choices to the BCI user in terms of specific end targets/destinations. Once the destination is selected by the BCI user, it can be reached autonomously in the unsupervised mode (cf. Figure 4.2(d)).

The number of commands needed to reach the specified destination is detailed in Table 6.3[3]. The commands required to reach 'Room 1' through the iAUI and the non-adaptive interface is 13. However, the iAUI requires only seven NC commands while the non-adaptive interface requires 13 NC commands. This results in a difference of duration of 6 trial times. This is the main advantage of the iAUI − it prioritizes the commands available to the user so that they do not need to perform NC, and can go to the next available choice. Figure 6.6 shows the mobile robot trail that reaches each of the three destinations by implementing the commands from the adaptive, non-adaptive and the autonomous interfaces, respectively. Table 6.4 details the mission time and the total cost incurred in completing a task, as suggested in reference [300]. The total cost is calculated from the nominal time (which is the minimum time required for the robot to reach the destination) and the mission time (the total time required to reach the destination) for each task. Both these terms are discussed later in this section. The total cost given in Table 6.4 is obtained by using the ideal TSD, i.e., the command is issued manually with 100% accuracy to represent an ideal situation. From Table 6.3 and Table 6.4 it is clear that the time and number of commands required to reach any of the specified locations are reduced if the interface is adaptive. This is compared with other interface design strategies (cf. Table 6.5) and is explained in the next section. It is important to note here that the path traversed when using both the adaptive and the non-adaptive approaches is the same. However, the path traversed when using the autonomous mode may be different, as the robot

---

[3]The methodology to calculate the values obtained in the table is explained in Appendix A.

TABLE 6.3 Number of Commands Required to Control the Robot in an Unstructured Environment Shown in Figure 6.4

Destination	No. of commands to reach the specified destination (with No-Control)		
	Adaptive interface	Non-adaptive interface	Autonomous interface
Room 1	13 / (07)	13 / (13)	01 / (13)
Room 2	09 / (07)	11 / (12)	01 / (17)
Cupboard	20 / (17)	22 / (20)	01 / (25)

Reproduced with permission from [338].

tries to reach the destination by an obstacle avoidance technique that is based on the minimum distance criterion. In this criterion, the shortest distance between the present location and the target location is considered without any previous knowledge of the whereabouts of the obstacles on that shortest path. Therefore, in the autonomous mode, the mobile robot may follow a different path to what a BCI user may choose, because the BCI user is aware of the entire robotic arena from the plan layout, including any obstacles, and hence might alter the travel path to avoid them.

A detailed list of the commands sent from the interface is given in Table 6.6. This list also includes the number of NC commands from the user. The data in this table can be used to make a very important observation: only one single NC or no-control command needs to be issued by the user when utilizing the adaptive interface (i.e., iAUI), for any of the three tasks. The same number without implementing the adaptive interface in place (i.e., using the fixed interface) is 20. A single NC is required when the user intends to select a task that is available in the second or subsequent available options within the interface. This suggests that the adaptive nature of the interface is very valuable in minimizing the time lost in issuing no-control commands. Thus, the probabilities that the tasks will be selected by the BCI user have certainly been appropriately calculated, and used to enhance the otherwise low communication bandwidth. In addition, as shown in Table 6.3 and Table 6.6, the total number of NCs required for all the three tasks is also reduced; from 45 (with the non-adaptive interface) to 31 (with the adaptive interface). The total number of commands required to complete all three tasks ('Room 1', 'Room 2' and 'Cupboard') has also reduced, from a likely number of 46 (with the non-adaptive interface) to 42 (with the adaptive interface). Both these reductions (in the actual number of commands and the number of NCs) contribute to making communication faster when using the adaptive interface.

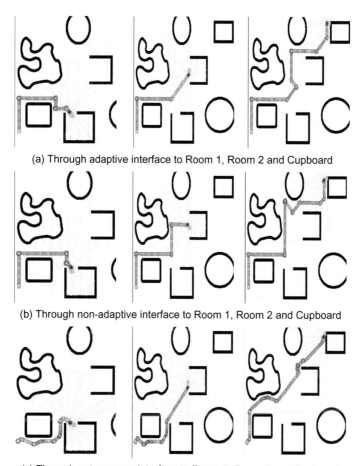

(a) Through adaptive interface to Room 1, Room 2 and Cupboard

(b) Through non-adaptive interface to Room 1, Room 2 and Cupboard

(c) Through autonomous interface to Room 1, Room 2 and Cupboard

FIGURE 6.6   Mobile robot trail for three destination targets using different interfaces. (a) Through the adaptive interface Room 1, Room 2 and Cupboard. (b) Through the non-adaptive interface to Room 1, Room 2 and Cupboard. (c) Through the autonomous interface to Room 1, Room 2 and Cupboard. *Partially reproduced with permission from [338]*

6.4.2 Evaluating the Interface Designs

6.4.2.1 *Evaluation Quantifiers*

This section first evaluates the different forms of the proposed interface on the basis of the total cost incurred in completing the three tasks, i.e., reaching 'Room 1', 'Room 2' and the 'Cupboard' from the original position, as shown in Figure 6.4. The total cost to accomplish a task with all three forms of the interface is also compared with the other interface designs discussed in references [298−300] using the same cost

TABLE 6.4 Performance Measure to Control the Robot in the Unstructured Environment Shown in Figure 6.4

Destination	BCR (for Room 1)			BCR (for Room 2)			BCR (for cupboard)		
	Adaptive interface	Non-adaptive interface	Autonomous interface	Adaptive interface	Non-adaptive interface	Autonomous interface	Adaptive interface	Non-adaptive interface	Autonomous interface
Nominal time	72	73	66	65	83	86	128	134	121
Mission time	100	130	70	80	115	90	185	210	130
Mission time ratio	1.38	1.78	1.06	1.23	1.38	1.04	1.44	1.56	1.07
Concentration time	52	52	4	36	44	4	80	88	4
Concentration time ratio	0.72	0.71	0.06	0.55	0.53	0.04	0.62	0.66	0.03
Total cost	2.1	2.49	1.12	1.78	1.91	1.08	2.06	2.22	1.10

Reproduced with permission from [338].

TABLE 6.5 Evaluation of Strategies to Control a Wheelchair by BCI

	Satti		Rebsamen		MAIA	Toyota	Iturrate	
	Self-paced	Synchronous	No false stop	Some false stops			Complex envt.	Open space
Number of false stops	NA	NA	0	1.21	NA	NA	NA	NA
Nominal time (s)	88	88	100	100	100	17	24	64
Mission time (s)	115.19	228	112	128	200	22.88	571	659
Mission time ratio	1.3	2.59	1.13	1.28	2	1.35	25	10.3
Concentration time (s)	15	96	12.6	28.3	200	22.88	447	439
Concentration time ratio	0.17	1.09	0.13	0.28	2	1.35	18.6	6.8
Total cost	1.47	3.68	1.26	1.56	4	2.7	43.6	17.1

Reproduced with permission from [338].

TABLE 6.6 Commands Sent to Control the Robot in the Unstructured Environment shown in Figure 6.4

Adaptive			Non-adaptive		
Room 1	Room 2	Cupboard	Room 1	Room 2	Cupboard
F	F	F	F	F	F
3 NC	3 NC	3 NC	3 NC	3 NC	3 NC
R	R	R	NC	NC	NC
F	F	F	R	R	R
3 NC	3 NC	3 NC	F	F	F
R	F	F	3 NC	3 NC	3 NC
F	L	L	F	F	F
L	F	F	NC	NC	NC
L	F	F	R	L	L
F	F	L	F	F	F
F	F	F	NC	F	3 NC
R	NC	R	L	NC	NC
F		F	F	R	R
L		3 NC	NC	NC	NC
F		R	L	R	R
NC		F	F	·F	F
		3 NC	NC	F	F
		F	R	NC	NC
		L	NC	NC	R
		F	L		F
		F	F		NC
		3 NC	NC		L
		F			F
		NC			NC
		L			R
		F			F
		NC			F
					NC
					L
					F
					NC
					L
					F
					NC
					R
					NC
13 / 7 NC	9 / 7 NC	20 / 17 NC	13 / 13 NC	11 / 12 NC	22 / 20 NC

criteria. The parameters detailed by Rebsamen et al. [300] are used to calculate the cost. They include calculating the mission time ratio and the concentration time ratio. Both these parameters can be obtained from the nominal time, the mission time and the concentration time for each task.

The nominal time is defined as the minimum time needed for the robot to reach the destination. The speed of the robot is 0.2 m/s and it covers a variable distance to reach each of the mentioned destinations. Hence, the nominal time has been calculated separately for each of the destinations and these are different for the supervised (adaptive and non-adaptive) and autonomous control interfaces because the travel path may vary with each interface.

The mission time is the time to select a destination plus the total traveling time [300]. Similarly, the concentration time for the application in reference [300] is the mission time minus the nominal time. However, the nature of the interface presented in this book is based on a synchronous MI BCI concept which is different from the P300 presented by Rebsamen et al. [300]. Therefore, the concentration time in the present work should be calculated differently. One possibility is to use the number of trials (by excluding the trials with no-control) in which the user issues commands by MI that is performed for 4 s in the trial duration[4]. The concentration time is the mission time minus the relaxation period. It is quite logical to assume NC as a relaxation period because the user is in a relaxed state and is not performing any mental imagination. Thus, since the number of commands for guiding the robot to 'Room 1' in the adaptive interface is 13, the concentration time is obtained as:

$$
\begin{aligned}
\text{Concentration Time} &= \text{number of MI commands} \\
&\quad \times \text{duration of MI in a single trial} \\
&= 13 \times 4 \\
&= 52 \text{ s}
\end{aligned}
$$

Similarly, with the autononous interface (cf. Figure 4.2(d)), the command to select the task 'Room 1' or 'Room 2' can be sent in the first trial, while the command to select 'Cupboard' can be sent in the second trial (albeit with one NC), thereby resulting in a concentration time of 4 s. The concentration time can be calculated in a similar way for the other interface design(s) for different destinations by utilizing the information in Table 6.3 and Table 6.6, as shown in Table 6.4. Appendix A discusses the

---

[4]The trial duration time may be 5 s or 7 s. However, the user performs actual MI for 4 s only and the command is sent after (trial duration − 1) seconds. The 1 s time period between issuing the BCI command and displaying the refreshed interface is necessary for communication between the controlled device and the interface.

calculation of the mission time, concentration time and the nominal time for the proposed interface(s) for one destination, 'Room 1'.

Therefore, the total cost to complete a specific task can be calculated as:

$$Total\ cost = concentration\ time\ ratio + mission\ time\ ratio \qquad (6.1)$$

Where the concentration time ratio is calculated as:

$$Concentration\ time\ ratio = \frac{concentration\ time}{nominal\ time} \qquad (6.2)$$

Here, the nominal time is the minimum amount of time to complete a given task, and is shown along with the concentration time for each of the tasks in Table 6.4.

The mission time ratio is calculated as:

$$Mission\ time\ ratio = \frac{mission\ time}{nominal\ time} \qquad (6.3)$$

Again, in the present interface design it is possible that the mission time may not be exactly the same as the one discussed in reference [300]. This interpretation (of mission time as the time to select a destination plus the total traveling time) is accurate in the proposed work only for the autonomous design where both these quantifiers (the time to select the destination and the time for mobile robot travel) are independent of each other. However, in the supervised (both fixed and adaptive) form of the interface, the user can issue a command while the mobile robot is in motion, i.e., there is an overlap in both these quantifiers. Therefore, the mission time in the present work is simply the total time required to reach the target destination from the original position, starting at the time at which the first command is initiated (including the first trial time). Table 6.4 lists the mission times for different tasks with each design.

With knowledge of the quantifiers discussed above and their values (from Table 6.4), it can be concluded that the total cost for maneuvering the mobile robot to the specified locations is lowest when using the autonomous interface (the average is 1.10). A second conclusion that can be drawn from the table is that the total cost for the adaptive interface, i.e., the iAUI design, is much less for all three task locations than that for the fixed/static interface design (the average is 1.98 [adaptive] vs. 2.21 [fixed]), but this is more than that for the autonomous interface (the average is 1.10). However, with an autonomous interface the user loses the freedom of specifying the desired robotic movement. Major requirements of autonomous designs are limited, predefined tasks for user selection and the need for a stored map of the robotic arena. In contrast, a major advantage of the adaptive interface over the autonomous interface is the freedom that the user has to select any desired task for the robotic movement. However, this incurs a much higher cognitive load. Therefore, some form of a combined approach may be a preferred method.

6.4.2.2 Comparing the Interfaces

As discussed before, the proposed interface(s) have also been compared with the existing approaches listed in Table 6.5 (reproduced from references [300] and [334]). The cost of the brain-controlled wheelchair (BCW) obtained by Rebsamen et al. [300] is 1.26 and 1.56, with no false stops and some false stops, respectively. The costs of synchronous and self-paced BCIs by Satti et al. [334] are 3.68 and 1.47, respectively. The costs of MAIA [282] and Toyota BCWs [299] are 4 and 2.7, respectively. Similarly the cost of Iturrate et al. [298] in complex and simple environments is 43.6 and 17.1, respectively. All these three approaches will be discussed briefly in this section.

The MAIA BCW proposed in reference [282] utilizes an MI-based BCI. The average mission time for this system ranged between 130 s and 270 s, therefore, the mean mission time was taken to be 200 s. The nominal time for this task was 100 s. The mission time ratio and the concentration time ratio obtained with this approach was shown to be 2, thereby resulting in a total cost of 4. A major limitation of this approach is the use of the orientation control strategy [282], which uses prior knowledge about the location of the target. The BCW proposed in this work reorients itself towards the target if it drifts away from the same by a certain threshold. Hence, this approach is not practical as it utilizes prior information about the destination and therefore does not offer real independence to the BCI user.

The Toyota wheelchair proposed in reference [299] uses an MI-based BCI system. The task is to maneuver the wheelchair along a path shaped in a figure of eight. One subject controlled the wheelchair in this study and completed the task in 22.88 s with a nominal time of 16.96 s. The mission time ratio for this system was 1.35. Moreover, the concentration time ratio for this system was also 1.35, thereby resulting in a total cost of 2.70.

Iturrate et al. [298] proposed a P300-based BCI and virtually reconstructed the surrounding environment using a laser range scanner. The user was presented with different points in the virtual image of the surrounding environment and these were selected using P300 signals. However, this method incurred a large cost in terms of both concentration time and mission time because the user had to concentrate every time the environment reconstruction occurred.

The BCW proposed by Rebsamen et al. [300] utilizes a P300-based BCI approach. In this method, only one selection is needed, at the beginning of the mission, which uses the P300 signal to select the desired location from a number of available options. Since this approach is based on the P300 and the user is required to make only one selection, the user concentration time is only necessary at the beginning of the first command issuance task. This BCW therefore has a very low concentration time ratio, which gives a very low control cost of 1.26 and 1.56, with no false stops and with some false stops, respectively. However, the user loses control of the BCW once the target location is selected. In addition, this method also requires prior knowledge of the environment in the form of a map containing information about the predefined end destinations, which is displayed on the P300-based user interface. This makes the BCI user dependent on the interface as well as on the predefined target destinations.

The synchronous and self-paced BCWs proposed by Satti et al. [334] have an average total cost of 3.68 and 1.47, respectively. However, these BCWs have the drawback that they do not have a command to start/ stop the robot once the desired target destination is reached. In addition, as has been discussed earlier, the BCW proposed in reference [334] continues to move in the forward direction (without any command from the user) until it reaches a junction, and the user can only control the left and the right movements of the robot. As explained in Section 2.10.1 (cf. Figure 2.17), these types of designs fail when the robot reaches a junction with three possible paths, or when the user has made a mistake and wishes to turn backward. Therefore, utilizing this concept in a real-time practical BCW may be difficult.

A major issue with all these designs is that they do not really provide the much-needed freedom of task selection because they are primarily autonomous movement-based control design concepts. In contrast, the interface design proposed in this book simultaneously provides the benefit of having completely autonomous and completely supervised modes, in addition to a combination of the two approaches, for device control. At the same time, compared to the existing designs the costs of the proposed approach are 1.98, 2.21 and 1.1 for adaptive, non-adaptive and autonomous interface designs, respectively (cf. Table 6.4). These costs are comparably low without really compromising on the control aspect. Of all the BCW methodologies discussed here, the autonomous interface-based design proposed in this book is thus the most cost-effective approach with an in-built obstacle avoidance strategy. However, its limitation is the need for a predefined map with a limited number of target destinations. In this respect, the adaptive interface-based design proposed in this book is a better choice, as it gives complete freedom in terms of supervisory control and it also incorporates a collision avoidance strategy. However, this approach is not as cost-effective as the autonomous interface or the one proposed by Rebsamen et al. [300]. Therefore, it is more practical and appropriate to utilize the autonomous and adaptive interfaces proposed in this book simultaneously. The coarse commands, such as entry to specific areas/rooms/broader destinations, can be managed by autonomous means while the finer movement commands (after entering specific areas/rooms) can be administered through the supervised adaptive interface (the iAUI). The approximate cost of using a combination of these interfaces will in practice depend on the individual application, but considering that the BCI user has the actual freedom to manage tasks at all times, this seems the best feasible option for all.

The next section discusses the real-time maneuvering of the mobile robot, using only the user's MI EEG signals, to the three locations identified in Figure 6.4 and labeled 'Room 1', 'Room 2' and 'Cupboard'.

6.5 MANEUVERING THE SIMULATED MOBILE ROBOT USING ONLY MI EEG

This section begins with an explanation of the training paradigm used in the BCI system to set up the classifier for online, real-time operation of the proposed robot control interface.

6.5.1 Training Paradigm

A paradigm similar to the one discussed in Chapter 4 is used to train the BCI users and to set up the classifier, so that the users are acquainted with how the actual interface works. This training paradigm is shown in Figure 6.7 and simply displays the left hand or the right hand arrow (marked in red). The trial time for displaying the arrows is

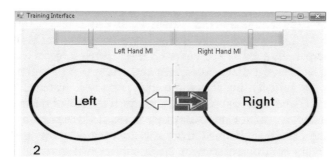

FIGURE 6.7   Training paradigm for the mobile robot control.

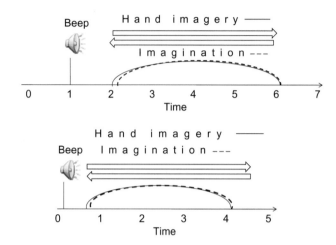

FIGURE 6.8   Trial time for the training paradigm.

shown in Figure 6.8. The right hand or left hand arrow is displayed continually for approximately 4 seconds as per the trial MI period, during which the user should imagine the specified task. If the trial duration chosen is 7 s, then the arrow is highlighted in red from 2 s until 6 s after the beep at 1 s. If the trial duration chosen is 5 s, then the arrow is highlighted from 0.5 s to 4.25 s after the beep at 0.25 s (cf. Figure 6.8). This task is repeated 60 times; i.e., there are 60 trials, with 30 right hand arrows and 30 left hand arrows being displayed randomly in every run. Before beginning the robot control paradigm on any given day, one set of EEG session training data, collected using the dry electrodes, is used to train the classifier after five-fold CV (cf. Section 2.9.1).

6.5.2 Methodology

The adaptive interface has been utilized to maneuver the robotic device to the three destinations marked 'Room 1', 'Room 2' and 'Cupboard' by using only MI EEG signals from five subjects (cf. video at [344]). All five subjects were male, aged 23−35 years. Of these, three subjects were experienced, one had intermediate experience and one was new to the area. The subjects were initially instructed on the proposed training paradigm (cf. Figure 6.7) and were asked to only perform the MI of a left hand and right hand movement for one run of 60 trials at the beginning of every session. The classifier obtained from this training set was used for the online mobility control experiment. Each subject was given a maximum duration of 12 minutes to reach any specific destination. If the subject did not reach the destination within this duration then the attempt was considered unsuccessful. In addition, if the subject did not feel at ease during a particular attempt (either due to tiredness during an attempt or being unable to concentrate after certain duration of time within the trial attempt) then the particular attempt was aborted and a new attempt was initiated.

6.5.3 Results and Discussion

Table 6.7 gives a list of the attempts made by all subjects to reach the three destinations. Subject K03 performed the experiment in five sessions on Days 1, 2, 3, 5 and 8, while subject I05 performed the experiment in two sessions on Day 1 and Day 10. The other subjects were

TABLE 6.7   Number of Attempts by All Subjects

Subj. V01	Session 1 Day 1	Session 2 Day 2
Room 1	—	1*
Room 2	—	1*
Cupboard	1*	

Subj. C02	Session 1 Day 1	Session 2 Day 2	Session 3 Day 3	Session 4 Day 4	Session 5 Day 5
Room 1	3*	—	—	—	—
Room 2	1	4	4	2*	—
Cupboard	—	—	—	1	1*

Subj. K03	Session 1 Day 1	Session 2 Day 2	Session 3 Day 3	Session 4 Day 5	Session 5 Day 8
Room 1	3	1	1*	—	—
Room 2	—	3	1	3	1*
Cupboard	—	—	1	1	1*

Subj. G04	Session 1 Day 1	Session 2 Day 2	Session 3 Day 4	Session 4 Day 5
Room 1	2	2	1	1*
Room 2	—	—	1	1*
Cupboard	—	—	—	1

Subj. I05	Session 1 Day 1	Session 2 Day 10
Room 1	2	1*
Room 2	—	1*
Cupboard	—	1

*Indicates successful attempt.

reasonably regular. Four subjects could reach one destination by the end of session 3 and two destinations in a single session. All subjects except one who reached a particular destination in a specific session did so at the first attempt. Moreover, by session 5, almost all the subjects had reached at least two destinations, and were thus well acquainted with the adaptive behavior of the interface.

The training accuracy and the kappa value for each of the subjects, along with the best model parameters obtained after five-fold CV from the training data, are shown in Figures 6.9 to 6.13 for all successful attempts to reach the specified destination. The figures also show the robot trail during the online task, in which the simulated mobile robot is maneuvered from the origin to the destination marked with an orange icon, and the user controls the robot movement solely by performing the MI in accordance with the iAUI. Table 6.8 shows the performance measures for all subjects while they try to complete the three tasks. The performance measure is calculated only for successful attempts, otherwise it is blank. The overall cost for maneuvering the mobile robot is higher than that obtained through the manipulated TSD shown in Table 6.4. This is as expected, because the error while issuing a specific command using the manipulated TSD is 0%. However, there is considerable error during the issuance of commands through user MI EEG signals, both from the BCI user and in signal processing. Since this occurs in real-time and knowledge of the user's class label is unavailable, it is not possible to measure the amount of error exactly.

The performance and the model parameters shown in Figure 6.9 for subject V01 are obtained for the training set of one run consisting of 60 trials with a trial time of 5 s (cf. training paradigm in Figures 6.7 and 6.8). The classifier obtained here was used for the online robot control task in the same session. This subject was the most experienced of all the subjects invited to take part in the experiment, and was able to reach all three destinations identified on the robot arena in the same session, after five unsuccessful attempts in earlier sessions.

For subject C02, one run of 60 trials was used to obtain the classifier using a trial time of 5 s (cf. training paradigm in Figures 6.7 and 6.8). This classifier was used for the robot control task in the same session. This subject was also experienced with the BCI system and reached Room 1 on the third attempt of the first session on Day 1. However, the subject felt exhausted and hence attempts to maneuver the robot to the other destinations were carried out in another session on the next day. The subject also needed more time to rest in between the trials, and hence the trial time was extended to 7 s (cf. Figure 6.8). Before beginning a session on any day, one run of 60 trials was used to obtain the classifier by using the training paradigm (cf. Figure 6.7), and subsequently the classifier was chosen for the robot control task. The subject

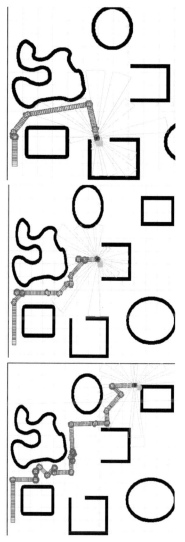

Frequency band*				Performance	
lf_1	hf_1	lf_2	hf_2	Accuracy	Kappa
8	10	20	24	86.66	0.73

*5s trial

Best model parameters and performance during training

Mobile robot trail for the three destination targets in the robotic arena

FIGURE 6.9   Best performance with subject V01 using the iAUI. Best model parameters and performance during training. Mobile robot trail for the three destination targets in the robotic arena. *Partially reproduced with permission from [338].*

reached the target destination 'Room 2' on the second attempt in the fourth session. The subject then also made one unsuccessful attempt to reach the third destination ('Room 3'). The same methodology was

Frequency band*				Performance	
lf_1	hf_1	lf_2	hf_2	Accuracy	Kappa
8	10	16	28	85.00	0.70

*5s trial

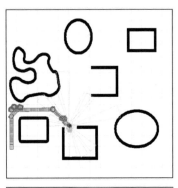

Frequency band				Performance	
lf_1	hf_1	lf_2	hf_2	Accuracy	Kappa
8	10	14	22	88.33	0.77

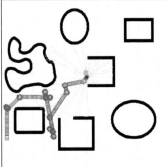

Frequency band				Performance	
lf_1	hf_1	lf_2	hf_2	Accuracy	Kappa
10	12	16	20	88.33	0.76

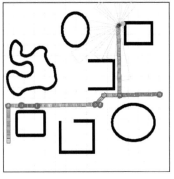

Best model parameters and performance during training   Mobile robot trail for the three destination targets in the robotic arena

FIGURE 6.10 Best performance with subject C02 using the iAUI. Best model parameters and performance during training. Mobile robot trail for the three destination targets in the robotic arena.

implemented the next day, and the subject reached 'Room 3' at the first attempt. The best attempts (and the performance/model parameters) for this subject are shown in Figure 6.10.

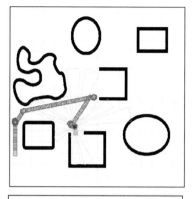

Frequency band				Performance	
lf_1	hf_1	lf_2	hf_2	Accuracy	Kappa
8	12	20	24	100.00	1.00

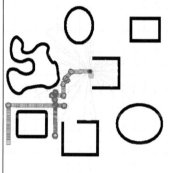

Frequency band				Performance	
lf_1	hf_1	lf_2	hf_2	Accuracy	Kappa
8	12	16	26	96.66	0.93

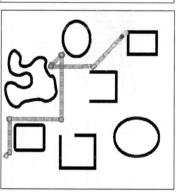

Best model parameters and performance during training

Mobile robot trail for the three destination targets in the robotic arena

FIGURE 6.11   Best performance with subject K03 using the iAUI. Best model parameters and performance during training. Mobile robot trail for the three destination targets in the robotic arena.

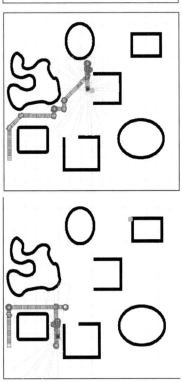

Frequency band				Performance	
lf_1	hf_1	lf_2	hf_2	Accuracy	Kappa
9	11	14	22	85.00	0.70

Best model parameters and performance during training

Mobile robot trail for the three destination targets in the robotic arena

FIGURE 6.12   Best performance with subject G04 using the iAUI. Best model parameters and performance during training. Mobile robot trail for the three destination targets in the robotic arena.

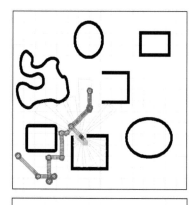

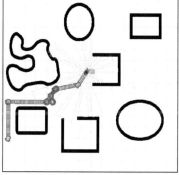

Frequency band				Performance	
lf_1	hf_1	lf_2	hf_2	Accuracy	Kappa
10	12	16	24	88.33	0.77

Best model parameters and performance during training	Mobile robot trail for the three destination targets in the robotic arena

FIGURE 6.13   Best performance with subject I05 using the iAUI.

Subject K03 was also an experienced subject. This subject preferred a trial time of 7 s and performed training on the paradigm for two successive days for about one hour each (including electrode mounting, etc.). The subject reached the destination marked 'Room 1' on the first

TABLE 6.8   Performance Evaluation for Real-Time Control of the Robot in the Unstructured Environment Shown in Figure 6.4

Parameter	Subjects (Room 1)				
	V01*	C02*	K03	G04	I05
Nominal time	72	72	72	72	72
Mission time	115	415	371	630	756
Mission time ratio	1.59	5.76	5.15	8.75	10.50
Concentration time	80	172	92	164	192
Concentration time ratio	1.11	2.38	1.27	2.27	2.66
Total Cost	**2.7**	**8.14**	**6.42**	**11.02**	**13.16**
Parameter	Subjects (Room 2)				
	V01*	C02	K03	G04	I05
Nominal time	65	65	65	65	65
Mission time	170	504	588	637	350
Mission time ratio	2.61	7.75	9.04	9.8	5.38
Concentration time	116	188	176	268	116
Concentration time ratio	1.78	2.89	2.70	4.12	1.78
Total Cost	**4.39**	**10.64**	**11.74**	**13.92**	**7.16**
Parameter	Subjects (Cupboard)				
	V01*	C02	K03	G04	I05
Nominal time	128	128	128	128	128
Mission time	515	560	539	DNR$	DNR$
Mission time ratio	4.02	4.37	4.21	–	–
Concentration time	276	112	156	–	–
Concentration time ratio	2.15	0.87	1.21	–	–
Total Cost	**6.17**	**5.24**	**5.42**	–	–

*5 s trial time elsewhere it is 7 s.
$DNR means Did Not Reach destination.
Partially reproduced with permission from [338].

attempt of the third session, on the third day of the experiment and then made one unsuccessful attempt to reach the destinations marked 'Room 2' and 'Cupboard'. The subject made another attempt (session 5) on Day 8 and reached 'Room 2' and the 'Cupboard' on the first attempt. The best attempts (and the performance/model parameters) of this subject are shown in Figure 6.11.

Subject G04 had intermediate levels of experience. This subject preferred a trial time of 7 s and made three unsuccessful attempts to reach 'Room 1' and 'Room 2'. However, in the fourth session on Day 5, the subject reached 'Room 1' and 'Room 2' at the first attempt and made one unsuccessful attempt to reach 'Cupboard'. The best attempts (and the performance/model parameters) for this subject are shown in Figure 6.12.

Subject I05 performed the BCI experiment for the first time. A similar methodology of one run of 60 trials was utilized to obtain the classifier parameters by using simple left hand and right hand MI. The subject was then instructed in the working of the robot paradigm later on this day, and so the robot maneuvering task was performed on the following day. This subject made two unsuccessful attempts to reach 'Room 1' in the first session on Day 1. This subject was not well, and hence only one more attempt was performed on Day 10 in the second session. The subject did one training session on this day and then was able to reach 'Room 1' and 'Room 2' on the first attempt. However, this subject could not reach the third destination 'Cupboard' on the first attempt of the second session. No further attempts were made after this. The best attempts (and the performance/model parameters) for this subject are shown in Figure 6.13.

6.5.4 Concluding Remarks

Certain points were discussed informally with the subjects in order to improve, enhance and benefit their control of the robot maneuvering tasks. Most of the subjects felt that the 'move backwards' command was of great help, primarily when an error led the robot to some wrong position. With just one selection in the form of a backward movement, the user can rotate the robot and start moving again. One subject felt that the 'Halt' command should be placed opposite the 'Forward' command, and that the 'Halt' command should preferably be located at the position with highest priority. However, if this is done, then other probable commands for maneuvering the robot would have to be placed at a lower priority, which may produce a sluggish system. At this time, the user was also advised that if a 'Forward' command is issued wrongly, then this can be partially overridden immediately in the next trial, either by issuing a 'Right' or a 'Left' command which would at the very least stop the forward movement of the robot. A user can also issue the 'Backward' command if it is prioritized at the first/second choice option to undo a wrong 'Forward' command. In addition, if the user fails to issue any of the above commands in the first two trials then the robot can be stopped using the 'Halt' command that is made available with the least priority. Lastly, if the user is unable to issue any of these commands in the complete cycle, then the robot will stop itself due to its inbuilt collision avoidance system, and it will also perform the usual stops at every junction (where the robot requires a user command for directing its movement from the junction). Thus, there is one option in every trial within the complete interface cycle to stop the forward movement of the robot. Subject K03 performed well in the training sessions of the experiment, but was not able to concentrate during the robotic maneuvering experiment because the robotic arena and the iAUI are displayed in different windows arranged side-by-side.

Overall, the iAUI system performed well with most subjects. However, the signal processing could improved by recognizing the NC state by using a one-versus-rest classifier rather than the simple one-versus-one classifier (that is used at present in combination with the post-processing module). Better recognition of the NC state is very important for controlling the interface. If a user-intended NC is not identified appropriately by the signal processing, then the user may inadvertently select a command to maneuver the robotic device. This aspect is discussed in the next chapter as part of the plan for the future work.

6.6 MANEUVERING THE PHYSICAL MOBILE ROBOT USING ONLY MI EEG

6.6.1 Methodology

This section discusses the implementation of the RQNN filtering technique and the iAUI for a real-time robot control application in a physical environment for the arenas shown in Figure 6.14 (cf. video in reference [344]). The EEG filtering is carried out by the proposed subject-specific RQNN model whose parameters are obtained by inner-outer five-fold cross-validation (cf. Section 5.4).

The commands are sent to the mobile robot through the user-centric iAUI module. Three subjects, all male and aged 21−35 years, attempted to maneuver the mobile robot in the physical arena through the online BCI system (cf. Figure 6.3). Of the three subjects, two were highly experienced while one had only limited experience. As discussed in the previous section, at the beginning of every session, the subjects performed MI of left and right hand movements in the training paradigm (cf. Figure 6.7) for a single run of 60 trials to set up the classifier for the online mobile robot navigation task. However, the subject-specific RQNN model parameters were obtained from the EEG data acquired during training on the first day of the experiment only (cf. Table 6.9 for subject-specific parameters obtained from the first session). This is

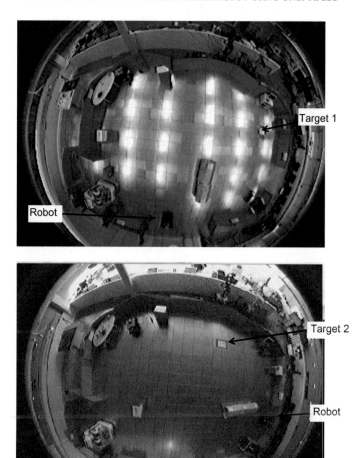

FIGURE 6.14   Screenshot of the robot arena (practical arena for 'Target 1' and 'Target 2').

TABLE 6.9   RQNN Parameters Using Inner-Outer Five-Fold CV (subjects in robotics lab.)

Subject	RQNN Parameters				Frequency Bands				Accuracy	Kappa
	β	m	ζ	β_d	lf_1	hf_1	lf_2	hf_2	(Training)	
V01	3.5	0.25	415	3	8	10	20	24	91.66	0.83
K03	6.0	0.35	815	4	10	12	14	16	95.00	0.90
S04	5.0	0.50	15	7	11	13	16	28	81.66	0.63

because the RQNN parameter selection approach is very time-consuming (cf. Section 5.5 and Section 7.2). Thus, the first stage of the inner-outer five-fold cross-validation to obtain subject-specific RQNN parameters was performed only once, whereas the second stage, which obtained subject-specific frequency bands, was performed prior to every session. The classifier obtained from the training paradigm was used for the online mobility control experiment. Each subject was allotted a maximum time of 12 minutes to reach the specific destination in the online experiment. This, as well as the other considerations, remains the same as with the robot control experiment performed in the simulated arena (cf. Section 6.5). The proposed RQNN methodology utilized in online EEG filtering for real-time MI-based robot control task using the iAUI is shown in the videos in reference [344].

6.6.2 Results and Discussion

Table 6.10 gives a list of the attempts made by all the subjects to maneuver the mobile robot to the two destinations marked 'Target 1' and 'Target 2' (cf. Figure 6.14). Table 6.11 gives the performance evaluation in terms of total cost, calculated from the concentration time and the mission time (cf. Section 2.12 and Section 6.4) for each subject while maneuvering the mobile robot to the specified target locations. The robot trail while reaching the destination is displayed in Figure 6.15.

Subject V01 was an experienced user and reached the destination 'Target 2' on the first attempt of session 1 (Day 1). However, V01 did not make more attempts to reach 'Target 1' on the same day. This subject then made three unsuccessful attempts to reach 'Target 1' in the second session on Day 9 before successfully reaching it on the fourth

TABLE 6.10   Number of Attempts by all Subjects

Subj. V01		Session 1 Day 1		Session 2 Day 9
Target 1		2*		—
Target 2		—		4*
Subj. K03	Session 1 Day 1	Session 2 Day 3	Session 3 Day 4	Session 4 Day 6
Target 1	2	3*	—	—
Target 2	—	—	3	2*
Subj. S04	Session 1 Day 1	Session 2 Day 3		Session 3 Day 4
Target 1	—	—		3
Target 2	2	2*		—

Indicates successful attempt.

TABLE 6.11   Performance Evaluation for Robot Control within the Arenas Shown in Figure 6.14

Parameter	Subjects (Target 1)		
	V01	**K03**	**S04**
Nominal time	37	37	DNR*
Mission time	77	315	
Mission time ratio	2.08	8.51	
Concentration time	40	64	
Concentration time ratio	1.08	1.72	
Total Cost	**3.16**	**10.23**	
Parameter	Subjects (Target 2)		
	V01	**K03**	**S04**
Nominal time	50	50	50
Mission time	140	462	441
Mission time ratio	2.8	9.24	8.82
Concentration time	68	144	140
Concentration time ratio	1.36	2.88	2.80
Total Cost	**4.16**	**12.12**	**11.62**

DNR = Did Not Reach.

attempt. The total costs for this subject for maneuvering the mobile robot to 'Target 1' and 'Target 2' were 3.16 and 4.16, respectively.

Subject K03 was also an experienced user. This subject reached the destination marked 'Target 1' on the fifth attempt of the second session (Day 3 of the experiment). K03 made three unsuccessful attempts (session 3) on Day 4 before successfully reaching the 'Target 2' on the second attempt on Day 5 (session 4). The unsuccessful attempts on Day 4, as discussed with the subject, were attributed to noticeable unavoidable sound disturbance on that day. The total costs for reaching the two targets 'Target 1' and 'Target 2' were 10.23 and 12.12, respectively.

Subject S04 had very limited experience. This subject reached the destination marked 'Target 2' on the second attempt of the second session

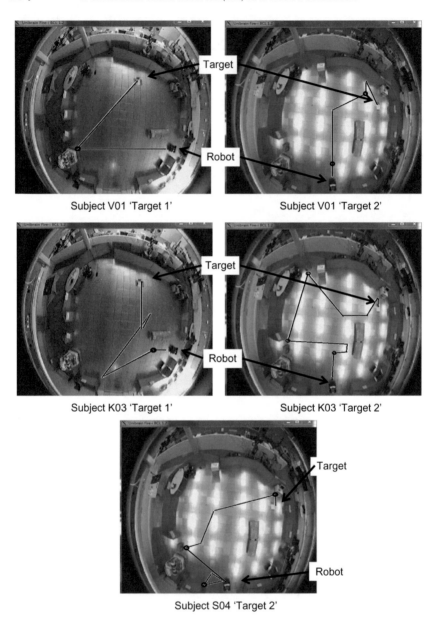

FIGURE 6.15   Physical robot trail with control through iAUI for each subject. *Partially reproduced with permission from [338].*

(Day 3 of the experiment). The total cost for reaching 'Target 2' was 11.62. The subject made three unsuccessful attempts (session 3) on Day 4 to reach 'Target 1'.

6.6.3 Concluding Remarks

In general, most of the participants were able to operate the adaptive interface and carry out the navigation task in the simulated arena as well as the physical robotic arena. Two of the three subjects who performed the experiment in the physical arena had also performed the experiment in the simulated environment. The subjects felt at ease during the initial training run and reached a higher level of training accuracy. However, when they began the real mobile robot maneuvering experiment in the physical arena, the feeling was different. Overall, these subjects, as discussed in informal conversation, felt a little more stressed while controlling the robot in the physical environment. This issue is of concern, as it can make the dynamics of the EEG obtained during the actual robot control experiment in the physical arena differ from those obtained during the training session. This can be because, although the EEG during the training session was also obtained in the physical arena, the added pressure of maneuvering the robots to a target location was absent. This may lead to wrong signal processing outcome during the online experiment that is ultimately responsible for controlling the mobile robot. However, this is due to the physical nature of the experiment rather than the design of the interface. The mission time ratio (cf. Table 6.11) increased by up to eight times the nominal time. This large increase is due to the BCI time required to develop the stimulation, recognize the desired command from the user as well as recover from BCI errors. However, the total cost for task completion is comparable to those obtained with other contemporary designs (cf. Table 6.4 and Table 6.11), and the interface presented here emphasizes the user having control, rather than the mobile robot.

In summary, these results indicate that the RQNN technique for EEG filtering and the adaptive nature of the interface for sending commands to the robot are suitable for controlling the mobile robot in a real-time online scenario within the simulated as well the physical robotic arena.

6.7 CONCLUSION

This chapter has discussed the real-time implementation of the proposed RQNN models for EEG signal enhancement, as well as novel interface designs for a mobile robot control task. The overall enhancement in terms of accuracy and kappa values with the RQNN filtered EEG is better than the unfiltered EEG for all the three subjects investigated in the BCI laboratory. In addition, it has been shown that the proposed interface designs (adaptive and autonomous) have the potential to provide true independence to the BCI user by implementing a combination of autonomous and adaptive designs, while the overall cost for the device control task increases by an acceptable extent. The interface is designed so that quicker maneuvering distant locations can be implemented by autonomous navigation, while more localized control can be managed through a supervised method. The overall costs in controlling the device are comparable to those obtained by Rebsamen et al. [300], and are much better than those obtained by MAIA, Toyota and Iturrate et al. [298]. However, as discussed in the previous section, there is substantial increase in the overall cost for completion of any task, which may be because of signal processing errors or user errors (i.e., wrong selections of tasks). Finally, the complete BCI system, including the subject-specific RQNN technique (for EEG filtering) and the user-centric iAUI (for enhancing the bandwidth) were implemented for real-time robot control in the physical environment. Thus, both techniques have been investigated in a real-time online application and are found appropriate for online applications.

CHAPTER

7

Conclusion

The major goal of this book was to address a number of important issues related to the complex, non-linear and non-stationary nature of the EEG signals in controlling robotic devices. To understand and define the issues and limitations associated with the development of a practical, efficient and accurate brain–computer interface (BCI) system, a detailed review of the BCI literature was carried out in, and discussed in Chapter 2. The overall focus of this review relates to the investigation of sensorimotor activities and a BCI system in general. The raw EEG signal, recorded non-invasively from the scalp of a user during motor imagery (MI), is intrinsically embedded within non-Gaussian noise, and the actual noise-free EEG has so far not been attained. Therefore, a detailed investigation into the use of quantum mechanics (QM) as a basis for filtering the EEG signals was undertaken, and presented in Chapter 3. A novel, QM-motivated, alternative, neural, information processing architecture using the Schrodinger wave equation (SWE) is proposed to filter and thereby enhance the information from the noisy input signal. This architecture, referred to as the Recurrent Quantum Neural Network (RQNN), is constructed using a layer of neurons within the neural network framework and recurrently computes a time-varying probability density function (*pdf*) for the measurement of the observed signal. The noisy input sample is encoded in terms of a particle-like wave packet that helps to enhance the signal using an unsupervised learning scheme without making any assumption about the signal type.

Once the quantum mechanics-based technique (i.e., the RQNN) was established to be suitable for filtering simple signals (cf. Chapter 3), it was then used in an evolving manner to filter the complex EEG signals (cf. Chapter 5). Genetic algorithm (GA), particle swarm optimization (PSO) and heuristics based approaches were initially investigated to appropriately tune/select the RQNN parameters and filter the EEG, but later a two-step, inner-outer, five-fold cross-validation (CV) technique was utilized for parameter selection. The RQNN-filtered EEG signal is more easily classified than that obtained by some other existing state-of-the art techniques. The QM-based RQNN technique can thus solve or at least partly solve the issues related to removing the noise embedded within the EEG. As discussed in Chapters 5 and 6, the RQNN-filtered EEG enhances the Classification Accuracy (CA) of the classifier, thereby potentially enhancing the overall performance of the BCI system.

However, to create a complete working BCI system that operates a robotic device, a major investigation in the field of interface design (cf. Section 2.10) was needed, to provide a communication link between the user's commands and the controlled device. A major challenge in two-class BCI systems is the inherent lower bandwidth of the communication channel, which may lead to a sluggish response in suitably controlling a mobile robotic device. Therefore, an intelligent and adaptive graphical user interface (GUI), which plays a very important role as a front-end display for the BCI user has been proposed (cf. Chapter 4). The framework of the proposed intelligent Adaptive User Interface (iAUI) is consistent with a range of applications, e.g., for controlling either a mobile robot or a robotic arm. The iAUI for a mobile robot offers a real-time, prioritized list of options for the user to choose from. This prioritized update of the iAUI is possible because information is obtained from the sonar sensors mounted on the mobile robot. The user can now make a quicker selection of a likely choice, thereby enhancing the speed of information transfer. Similarly, the interface for controlling the robot arm displays the list of available objects for user selection depending on the real-time information sent from the camera view of the robot arm. This novel, user-centered, adaptive interface design helps the BCI user to issue commands more quickly to the mobile device. To achieve further superior control over the device, work was also carried out on the post-processing stage of the BCI, i.e., on the classifier output or the Time-Varying Signed Distance (TSD) (cf. Section 2.8 and Section 4.3). In the following sections, the main contributions of this research are summarized.

7.1 CONTRIBUTIONS OF THE BOOK

7.1.1 Investigation of QM and SWE for Filter Development

The theoretical concepts from QM were investigated to develop the idea and propose the architecture for two different RQNN models for EEG signal enhancement in a BCI system. The RQNN models were constructed using a layer of neurons within the neural network framework. The traditional RQNN model is stimulated with an error signal, while the revised RQNN model is stimulated directly by the noisy input signal. The noisy input signal fed to the RQNN models is treated as a wave packet which recurrently evolves under the influence of the SWE and an appropriate set of learning rules. The potential function defines the force field within which the particles move and also directs the evolution of the wave packet. The expected value of the true input signal is calculated from the probability density function (*pdf*), which is the squared-modulus of the wave packets. The *pdf* evolves from an initial wave packet and learns the dynamics of the signal in an unsupervised manner without having any *a priori* knowledge of the signal or the noise. To the best of the author's knowledge, such an implementation of any quantum mechanics-based conceptual approach to EEG signal enhancement in BCI systems has not been undertaken before.

7.1.2 Understanding the Parameters of the RQNN Models

The evolution of the wave packets to produce desired results depends on the choice of the parameters of the RQNN model. The initialization of the wave packet, selection of values for the parameters in order to solve the SWE and the size of the potential well in order to localize and propagate the wave packet (this also represents the amplitude of the incoming noisy signal) are crucial and have been discussed in detail in Chapter 3.

7.1.3 Tuning/Selecting RQNN Model Parameters

Selecting and tuning the parameters of the RQNN models is important as they dictate the evolution/propagation of the wave packet that can ultimately give desired results. RQNN-based techniques have shown effectiveness in filtering the noise embedded within simple signals such as DC and sinusoidal waveforms (cf. Chapter 3). A PSO technique was employed to tune the RQNN model parameters to suit the filtering of these simple signals. The proposed model gives significantly lower root mean square errors (RMSE) than the Kalman filter for simple example signals. An inner-outer five-fold cross-validation technique is later employed to select the RQNN model parameters for enhancing complex EEG signals. Thus, a subject-specific RQNN model parameter can be derived (cf. Chapter 5), which results in better performance across multiple evaluation sessions.

7.1.4 Real-Time Implementation of the RQNN Model for EEG Signal Enhancement

The major advantage of the RQNN models is that they are completely data-driven and hence generic. It can be concluded from the discussion in Chapters 3 and 5 that the revised RQNN model is more robust than the traditional model, and provides accurate filtering even with inaccurately selected model parameters. Therefore, it can be said that the RQNN-based filter falls into the category of 'intelligent' or 'smart' filters primarily because of very little or no requirement of any kind of *a priori* information about the signal or the noise.

The real-time implementation of RQNN models as signal enhancement techniques for non-linear and non-stationary EEG signals is an important contribution of this research. Each incoming sample of the noisy EEG signal is treated as a wave packet. The wave packet evolves by learning the dynamics of the signal in an unsupervised manner without any *a priori* knowledge of the signal or the noise. These RQNN models have been applied using different learning algorithms and investigated using the BCI competition EEG datasets as well as the real-time data acquired from subjects in the BCI laboratory at the Intelligent Systems Research Centre (ISRC). The results (cf. Chapters 5 and 6) show that the performance of the classifier during the training and evaluation stages with the subject-specific RQNN model is enhanced compared to that using the raw EEG, Savitzky-Golay (SG) filtered EEG or even raw EEG with the power spectral density or the Bispectrum-based features. To the best of the author's knowledge, a real-time implementation and an on-line application of any quantum mechanics-based conceptual approach in BCI has never been undertaken before. This is a major contribution to the field of EEG signal filtering/separability enhancement [36,37,213] and has been explained in detail in this book.

7.1.5 Investigation into GUIs for Use in BCI Systems

The GUI plays a very important role as a front-end display for the BCI user. An investigation into different interface designs and shared control approaches was carried out, and discussed in Chapter 2. The study suggested that two-class GUIs having inherently higher accuracy characteristics can be utilized to maneuver/control a robotic device/ wheelchair. However, a major challenge in two-class BCI systems is the inherently low bandwidth of the communication channel, which may lead to a slower, sluggish response when controlling such devices. This investigation thus led to the development of the proposed interface designs for mobility control and arm control applications. The strength of the proposed interfaces is the overall framework that remains consistent for controlling the device, which could be a mobile robot, a robot arm or a wheelchair.

7.1.6 Intelligent Adaptive User Interface (iAUI) for Mobility Control

A very important aspect of BCI in the form of adaptive and user-centric interface designs has been proposed for mobility control applications in this book.

- The proposed interfaces are designed not only to provide feedback to the user, but also to enhance the overall communication aspect in a time-critical BCI application because they adapt to the changing environment, and thereby allow the BCI user to communicate rapidly in selecting up to six different options.
- This book has proposed an adaptive design that positions the choices within the interface dynamically (referred to as iAUI in Chapter 4), so that the most anticipated choices from the user are displayed in the most accessible places (based on the feedback from the controlled robotic device which sends sonar sensor information to the interface). It has been shown that the adaptive updating of the interface results in a reduced mission time and also in reduced overall cost for task completion (cf. Chapter 6).
- The interfaces are consistent because of the requirement to perform only one of the two mental imageries. This results in a MI BCI design with minimal error probability (two class 50%). However, the BCI user needs to be aware that the specific MI should be in agreement with the desired task on the monitor module of the interface. The user is also able to view the live video streamed from the camera of the pioneer robot. Hence, the scene around the controlled device is made available to the user (cf. Chapter 4). The iAUI design for mobility control has been evaluated practically with five subjects in a real-world environment and with three subjects in the physical environment. Is has also been compared with existing state-of-the-art approaches in the field of BCI (cf. Chapter 6). Thus, the concept of adaptive shared control for mobility has been successfully implemented, evaluated and compared with existing state-of-the-art methodologies in this book, and has been found not only to provide the usual feedback but also to be information-enhancing.

To the best of the author's knowledge, an adaptive and user-centric interface using similar strategies/concepts has not been undertaken or implemented before.

7.1.7 Adaptive User Interface for Robot Arm Control

The framework of the proposed interface remains consistent for controlling a robot arm for pick-and-place tasks.

- The interface for the robot arm control application updates itself dynamically depending on the list of available object choices identified within the camera view of the robot arm.
- The user is also able to view the live video that is streamed from the Kinect camera mounted on the end-effector of the Schunk robot arm [336] (cf. Chapter 4). Hence, the list of the objects visible from the robot arm camera is made available to the BCI user. However, the ARM interface has not been evaluated practically in real time with different subjects.

7.2 FUTURE RESEARCH DIRECTIONS

7.2.1 Tuning/Selecting the RQNN Model Parameters

The fixed and variable sets of parameters of the RQNN model for EEG signal enhancement are obtained in this study by using a heuristic approach as well as the two-step, inner-outer, five-fold CV (the first step being parameter selection and the second being frequency band selection) technique. However, as discussed in Section 5.4, this approach only gives the best combination of the four variable RQNN parameters and the four frequency bands from within a coarse range of combinations of the parameters and hence may not necessarily be the optimum one. Therefore, future work should obtain all the subject-specific parameters of the RQNN model (i.e., all nine parameters mentioned in Tables 5.14 and 5.15) and the four frequency bands while using an optimization technique such as PSO or GA within a single stage of the global search optimization. Thus, both the RQNN parameters and frequency band selection must be implemented within a single step. However, this would result in a huge multi-dimensional search space spanning 13 variables. To illustrate the extent of the problem, the two-step parameter optimization for each subject discussed in Chapter 5 and implemented on an 84-core set-up took approximately 24 hours of computational time. Thus, incorporating a global optimization will be a challenging task and will require appropriate prior optimization in terms of narrowing the regions of the search space for each variable.

7.2.2 Three-Class Classifier

The interface presented in this book selects a particular task by using both MIs. In addition, if neither MI is performed, then a no-control (NC) is assumed. A major issue that BCI users face in the present form of the interface is the inability to appropriately maintain the NC state. At this point, the interface seems to require three classes; each of the two Motor

Imageries (MIs) and NC. Therefore, the existing two-class Linear Discriminant Classifier (LDA) approach with post-processing can be replaced with a three-class LDA classifier. Thus, instead of the usual one-versus-one classifier to identify a particular MI for a pair of classifiers, a one-versus-rest or one-versus-all classifier can be used to distinguish a specific MI from a combination of the other MI and the NC state. A similar approach has been investigated by Geng et al. [226,347], who investigated a one-versus-all LDA classifier. Therefore, an analogous method with three output classes can be expected to identify their individual presence with more accuracy, and thereby the selection of an appropriate choice within the present design of the proposed interface can be performed more effectively. This requires further investigation into the classifier stage within the BCI system and is one of the priorities in the near future.

7.2.3 Hybrid BCI Systems

A hybrid BCI (hBCI) [348] is a strategy that utilizes more than one technique to operate a BCI. Conventionally, BCI systems are operated using MI, SSVEP, P300, ErrP, etc., but other biosignals such as electromyography (EMG)[1], electrocardiography (ECG)[2], eye gaze and many other approaches can also be used as complementary modalities to the conventional BCI techniques. Combining some of these techniques should take advantage of the positive features of the individual approaches and in the process enhance the overall performance of the BCI. The hBCI technique has been implemented in references [348–353], and is known to give satisfactory results for most users. A typical hBCI thus utilizes the advantages of different approaches and simultaneously does not put any additional burden on the BCI user. It is normally composed of one conventional BCI system and another system (which might be another conventional BCI system or one of the other techniques mentioned above), and thereby achieves specific goals in terms of a better CA, better control of the system or a reduction in the amount of attention required by the user. Thus, there is a major advantage in utilizing such hBCI technologies.

Simmons et al. [354] mentions that 40% of BCI users are not able to perform MI-related tasks. A similar study using a MI BCI by Guger et al. [355] shows that only 19% of the subjects are able to achieve an accuracy of 80–100%. Similarly, about 89% of the subjects are able to achieve an accuracy of 80–100% while using the P300-based BCI [355]. Thus, a combination of these approaches will enhance the overall

---

[1]Electromyography (EMG) is a technique for evaluating and recording electrical activity from the skeletal muscles.

[2]Electrocardiography (ECG) is a technique to measure the electrical activity of the heart.

performance of the BCI system by picking the technique that is most suitable to a particular user.

hBCIs can also use a combination of EEG and other brain imaging techniques, such as MEG, fMRI, NIRS, etc. Some hBCI techniques implemented by various groups are discussed in reference [356]. A hBCI combining MI-based BCI with ErrP has been successfully implemented for the detection and correction of false mental commands in reference [357]. Offline studies of a combination of MI with SSVEPs have also been carried out [349,358], and in reference [353], a hybrid BCI system was created in which a battery of flickering lights (SSVEP-BCI) was switched on or off by using a brain switch based on post-imagery beta event-related synchronization (ERS) [359].

Some hBCIs also utilize a combination of brain signals and other biosignals. A hBCI that combined SSVEP and heart rate variation (for an on/off switch) has been implemented in reference [350]. Such combinations give the added freedom to the BCI user of using the BCI only when they want to use it. Some other approaches which combine two completely different modalities, such as EMG and EEG activity [348,351] or EEG and eye gaze [360], have also been investigated. These techniques help in reducing the load of 'only performing the mental imagination' on the BCI user. A very good review of preliminary attempts and feasibility studies for developing different combinations of hBCIs is given in reference [352].

7.2.3.1 Hybrid BCI (SSVEP + ERD/ERS)

The interface presented in this book selects a task by associating an MI with the desired task icon on the selection bar. The interface in its present form is thus concerned with the usual left hand/right hand MI. However, the interface can also be modified to make it harmonious with a hybrid BCI by utilizing the concepts of SSVEP, similarly to those discussed in reference [349]. In this study, the authors validated the concept of hybrid BCI by utilizing a combination of MI and SSVEP. This concept can be accomplished as an extension in the proposed work by switching the task icons on the interface at specific frequencies, as expected for an SSVEP-based BCI system. The user is then expected to perform the usual MI of the left/right hand while focusing on the particular task icon choices. Thus, a combined SSVEP (through user focus) and MI (through ERD/ERS) based scheme can be utilized to identify the choice of the BCI user. This may ultimately lead to better accuracy in identifying a particular selection.

In the present book, the author has also proposed an interface to incorporate the hBCI technique by having control inputs from the EEG and the eye tracker device. The eye tracker device is used to locate the task that the user desires to perform (by moving the cursor over the task icon) while the actual selection or conveying the command to the device is implemented through the EEG command (when the user performs one of the pre-designated MI). This is discussed next.

7.2.3.2 Hybrid BCI (EEG + Eye Tracker System)

Future work could implement a hybrid BCI system by combining an eye tracker system and an EEG-based BCI to control a wheelchair or robotic device. A similar system has been investigated in reference [361], in which an eye gaze input is combined with BCI for human—computer interaction. The idea planned by this author is to highlight the user's initial intention to select specific commands through the eye tracker system (when the mouse cursor hovers over a specific task) while the actual selection (double clicking) of the command icon can be implemented using a single MI EEG.

The primary reason for combining EEG and an eye tracker system is that an eye tracker system alone can only provide psychophysical data in terms of the eye gaze location. Correspondingly, EEG measurement only provides information about the neuronal activity and none about the eye gaze location. Both EEG and eye gaze are attention-specific, but they fall into two separate categories, namely overt and covert attention [362]. Overt attention is inferred as the location a person is seeing. This can be interpreted as knowing where a person is looking through the movement of their eyes. However, one can also pretend to look at a location while simultaneously attending to another one. Attention is a mental procedure that uses neural processes, and is referred to as covert attention. EEG measurement can be utilized to detect covert attention while an eye tracking system can be utilized for overt attention. Both attention and eye movements are independent [363]. Thus, a scheme that simultaneously records eye gaze and neuronal activity could enhance the information available for controlling complex robotic movement tasks efficiently.

Figure 7.1 shows a multi-modal interface, in which the BCI user is expected to move the mouse cursor over any of the command icons by using the eye tracker device. Once the user is able to position the mouse cursor over the command icon, the next task is to perform any MI (which should be decided prior to the experiment in order to train the classifier) to actually select the command in lieu of the usual double clicking. A simple EEG-based BCI system has several drawbacks including lower accuracy and reduced bandwidth. Similarly, an inherent disadvantage of an eye tracker device is the issuance of false commands when the user blinks, which is a natural process but becomes a potential artifact in such systems. The anticipated advantage of the proposed hybrid system is that the number of false commands that are normally produced by BCI signal processing or the eye tracker device can be minimized. Another expected major advantage is that the eye tracker device

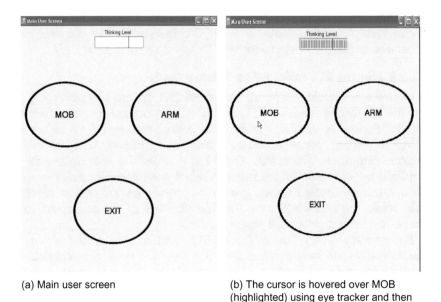

(a) Main user screen

(b) The cursor is hovered over MOB (highlighted) using eye tracker and then selected through the user's MI EEG.

FIGURE 7.1   Main interface with multi-modal input (EEG + eye tracker system). (a) Main user screen. (b) The cursor is moved over MOB (highlighted) using the eye tracker, and then selected through the user's MI EEG.

should reach the desired task icons much faster than the EEG synchronous BCI system concept. Therefore, the primary advantage expected from such a hybrid system is the gain in communication bandwidth and the reduction in false positives. The user would not be expected to wait for one or more trial durations (as is the case with synchronous paradigms) in order to select choices that are available at the second or later sub-stages within an interface. In fact, with a hybrid BCI, the user can quickly hover the mouse cursor above any of the desired command icons. Simultaneously, since this hybrid system only incorporates detection of just one mental imagination (or MI) versus the rest period, the probability of issuing wrong commands through the EEG is reduced. BCI signal processing only needs to identify whether the BCI user is performing any imagination or not, without going into the two-class or multi-class separation issues and related problems.

Thus, the problem can now be perceived as that of using a brain switch for a BCI end application task [78,352,353,359,364]. In addition, even if the BCI signal processing outcome sends a false command in the form of detecting a user imagination that did not in fact occur, in practice the command will only have an effect if the mouse cursor is located on any of the command icons. If the mouse cursor is not within the vicinity of any of the command icons, even a wrong command will have no effect on the outcome of the hybrid system. Thus, such a hybrid system is subject to fewer false positives, which has the potential to increase the communication bandwidth and the speed of information transfer.

The complete procedure of the proposed hybrid BCI system can be understood in terms of the interface shown in Figures 7.1, 7.2 and 7.3. Figure 7.1(a) shows the main user interface, in which the user is expected to move the mouse cursor over the command icon MOB by using the eye tracker system, and thus show the intention of selecting that task. In Figure 7.1(b), the user has hovered the mouse cursor over the command icon MOB and then performed a MI via EEG to select this command. At this point, the mobile robot control interface looks like Figure 7.2. If the user intends to send the command 'Forward' to the device, the cursor should first be moved to the command icon 'Forward' by using the eye tracker system, and then the selection is made by MI.

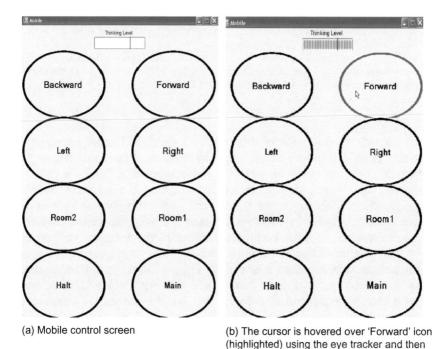

(a) Mobile control screen

(b) The cursor is hovered over 'Forward' icon (highlighted) using the eye tracker and then selected through the MI EEG

FIGURE 7.2 Mobile control interface with multi-modal input (EEG + eye tracker system). (a) Mobile control screen. (b) The cursor is moved over the 'Forward' icon (highlighted) using the eye tracker and then selected by using MI EEG.

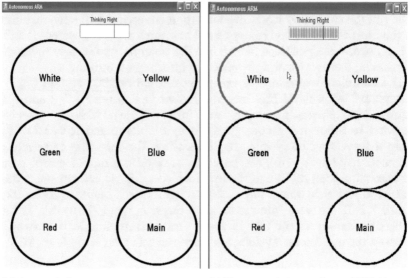

(a) Arm control screen

(b) The cursor is hovered over 'White' icon object (highlighted) using the eye tracker and then selected through the MI EEG

FIGURE 7.3   Arm control interface with multi-modal input (EEG + eye tracker system). (a) Arm control screen. (b) The cursor is moved over the 'White' icon object (highlighted) using the eye tracker and then selected by using MI EEG.

An important advantage of this type of hybrid BCI robot interface is that the tasks in the form of end destinations (such as 'Room 1', 'Room 2', etc.) as well as specific movement commands (such as 'Forward', 'Backward', etc.) can be made available for selection through a single interface. This saves the usual back and forth time in switching between sub-interfaces. In addition, the user is always in control and has complete freedom to issue commands either in the form of destination choices or specific movement commands.

The arm control interface is shown in Figure 7.3. As usual, the user is expected to move the mouse cursor via the eye tracker device, and hover it above the desired object icon that he/she wishes the robot arm to pick up. Once the command icon is highlighted, the user performs MI for the final selection (cf. Figure 7.3(b), in which the user selects the 'White' icon object).

The interface presented here maintains a similar usability procedure for controlling both the mobile robot and the robot arm. It also (based on the hybrid BCI concept) has the potential to truly give the BCI user the necessary freedom and control.

7.3 CONCLUSION

The main contributions of the book have been summarized in this chapter. It has been shown that two major contributions have been made and each of these has the potential to enhance the overall performance of the BCI system significantly. The significance of both these techniques has been justified. It can be safely said that the proposed, novel, QM-based methods for EEG signal enhancement, post-processing for synchronous BCIs and the design of a user-centric iAUI strategy for a safe and real-world applicable approach have addressed some of the major current issues and limitations in BCIs. Future work has also been proposed which would investigate the multi-dimensional parameter tuning approach using PSO or GA to simultaneously tune all the RQNN model parameters. Further work could also improve the proposed adaptive interface design by merging it with the hybrid BCI concept. It is expected that the methods and strategies developed in this book and the proposed future work will have a significant impact on the development of practical and adaptive BCIs.

Appendix A: Understanding Evaluation Quantifiers for the Proposed Interface

This purpose of this appendix is to explain how the evaluation quantifiers mentioned in Section 6.4 are calculated using the example shown in Figures 6.4 and 6.6, in which the robot is maneuvered to 'Room 1' using the proposed interface in the adaptive, non-adaptive and autonomous modes.

The first column in the table indicates time. Since the trial time is 5 s, the second column indicates the trial number at every 5 s. The mission time can be calculated as (number of trials × trial time), or simply the time at the end of the task completion or the end of the column 1. The third column is used to calculate the concentration time. In this column, a 1 indicates that the BCI user performs an MI, and a 0 shows that the user is in a relaxed state. Therefore, a summation of this column indicates the total concentration time of the BCI user. The nominal time is calculated from the fourth column which indicates a 1 if the robot is in motion and a 0 if it is stationary. The fifth or the last column indicates the task that is issued by the BCI user and is only used as a reference here. Tables A.1, A.2 and A.3 list the above-mentioned parameters for maneuvering the robot to 'Room 1' using the adaptive, non-adaptive and the autonomous interfaces, respectively.

TABLE A.1 Understanding the Evaluation Quantifiers to Reach 'Room 1' (Adaptive Interface)

Time	Trial	MI (1)/ Relax (0)	Robot moving (1)/ Stationary (0)	Command
1		1	0	
2		1	0	
3	1	1	0	F through R
4		1	0	
5		0	1	

(*Continued*)

TABLE A.1   (Continued)

Time	Trial	MI (1)/ Relax (0)	Robot moving (1)/ Stationary (0)	Command
6		0	1	
7		0	1	
8	2	0	1	NC
9		0	1	
10		0	1	
11		0	1	
12		0	1	
13	3	0	1	NC
14		0	1	
15		0	1	
16		0	1	
17		0	1	
18	4	0	1	NC
19		0	1	
20		0	1	
21		1	0	
22		1	0	
23	5	1	0	R through L
24		1	0	
25		0	1	
26		1	1	
27		1	1	
28	6	1	1	F through R
29		1	0	
30		0	1	
31		0	1	
32		0	1	
33	7	0	1	NC
34		0	1	
35		0	1	

(Continued)

TABLE A.1   (Continued)

Time	Trial	MI (1)/ Relax (0)	Robot moving (1)/ Stationary (0)	Command
36		0	1	
37		0	1	
38	8	0	1	NC
39		0	1	
40		0	1	
41		0	1	
42		0	1	
43	9	0	1	NC
44		0	1	
45		0	1	
46		1	1	
47		1	1	
48	10	1	1	R through L
49		1	0	
50		0	1	
51		1	1	
52		1	1	
53	11	1	1	F through R
54		1	0	
55		0	1	
56		1	1	
57		1	1	
58	12	1	1	L through L
59		1	1	
60		0	1	
61		1	1	
62		1	1	
63	13	1	0	L through L
64		1	0	
65		0	1	

(*Continued*)

TABLE A.1 (Continued)

Time	Trial	MI (1)/ Relax (0)	Robot moving (1)/ Stationary (0)	Command
66		1	1	
67		1	1	
68	14	1	0	F through R
69		1	0	
70		0	1	
71		1	1	
72		1	1	
73	15	1	1	F through R
74		1	1	
75		0	1	
76		1	0	
77		1	0	
78	16	1	0	R through L
79		1	0	
80		0	1	
81		1	1	
82		1	1	
83	17	1	1	F through R
84		1	0	
85		0	0	
86		1	0	
87		1	0	
88	18	1	0	L through L
89		1	0	
90		0	1	
91		1	1	
92		1	1	
93	19	1	0	F through R

(*Continued*)

TABLE A.1   (Continued)

Time	Trial	MI (1)/ Relax (0)	Robot moving (1)/ Stationary (0)	Command
94		1	0	
95		0	1	
96		0	1	
97		0	1	
98	20	0	1	NC
99		0	1	
100		0	0	
100	—	52	72	13 (07)

TABLE A.2   Understanding the Evaluation Quantifiers to Reach 'Room 1' (non-adaptive interface)

Time	Trial	MI (1)/ Relax (0)	Robot moving (1)/ Stationary (0)	Command
1		1	0	
2		1	0	
3	1	1	0	F through R
4		1	0	
5		0	1	
6		0	1	
7		0	1	
8	2	0	1	NC
9		0	1	
10		0	1	
11		0	1	
12		0	1	
13	3	0	1	NC
14		0	1	
15		0	1	

(*Continued*)

TABLE A.2 (Continued)

Time	Trial	MI (1)/ Relax (0)	Robot moving (1)/ Stationary (0)	Command
16		0	1	
17		0	1	
18	4	0	1	NC
19		0	1	
20		0	1	
21		0	0	
22		0	0	
23	5	0	0	NC
24		0	0	
25		0	0	
26		1	0	
27		1	0	
28	6	1	0	R through R
29		1	0	
30		0	1	
31		1	1	
32		1	1	
33	7	1	1	F through R
34		1	0	
35		0	1	
36		0	1	
37		0	1	
38	8	0	1	NC
39		0	1	
40		0	1	
41		0	1	
42		0	1	
43	9	0	1	NC
44		0	1	
45		0	1	

(Continued)

TABLE A.2   (Continued)

Time	Trial	MI (1)/ Relax (0)	Robot moving (1)/ Stationary (0)	Command
46		0	1	
47		0	1	
48	10	0	1	NC
49		0	1	
50		0	1	
51		1	1	
52		1	1	
53	11	1	1	F through R
54		1	0	
55		0	1	
56		0	1	
57		0	1	
58	12	0	1	NC
59		0	1	
60		0	1	
61		1	0	
62		1	0	
63	13	1	0	R through R
64		1	0	
65		0	1	
66		1	1	
67		1	1	
68	14	1	0	F through R
69		1	0	
70		0	1	
71		0	1	
72		0	1	
73	15	0	1	NC
74		0	1	
75		0	0	

(Continued)

TABLE A.2   (Continued)

Time	Trial	MI (1)/ Relax (0)	Robot moving (1)/ Stationary (0)	Command
76		1	0	
77		1	0	
78	16	1	0	L through L
79		1	0	
80		0	1	
81		1	1	
82		1	1	
83	17	1	0	F through R
84		1	0	
85		0	1	
86		0	0	
87		0	0	
88	18	0	0	NC
89		0	0	
90		0	0	
91		1	0	
92		1	0	
93	19	1	0	L through L
94		1	0	
95		0	1	
96		1	1	
97		1	1	
98	20	1	0	F through R
99		1	0	
100		0	1	
101		0	0	
102		0	0	
103	21	0	0	NC
104		0	0	
105		0	0	

(Continued)

TABLE A.2 (Continued)

Time	Trial	MI (1)/ Relax (0)	Robot moving (1)/ Stationary (0)	Command
106		1	0	
107		1	0	
108	22	1	0	R through R
109		1	0	
110		0	1	
111		0	1	
112		0	1	
113	23	0	1	NC
114		0	0	
115		0	0	
116		1	0	
117		1	0	
118	24	1	0	L through L
119		1	0	
120		0	1	
121		1	1	
122		1	1	
123	25	1	0	F through R
124		1	0	
125		0	1	
126		0	1	
127		0	1	
128	26	0	1	NC
129		0	1	
130		0	0	
130	–	52	73	13 (13)

TABLE A.3   Understanding the Evaluation Quantifiers to Reach 'Room 1' (autonomous interface)

Time	Trial	MI (1)/ Relax (0)	Robot moving (1)/ Stationary (0)	Command
1		1	0	
2		1	0	
3	1	1	0	R1
4		1	0	
5		0	1	
6		0	1	
7		0	1	
–	2–14	–	–	NC
–		–	–	
70		0	1	
70	–	4	66	1 (13)

Bibliography

[1] M.A. Hofman, D. Falk, Evolution of the Primate Brain: From Neuron to Behavior, Elsevier Science, 2012.

[2] G. Buzsaki, Rhythms of the Brain, Oxford University Press, 2006.

[3] R. Melillo, G. Leisman, Neurobehavioral Disorders of Childhood: An Evolutionary Perspective, Springer, 2004.

[4] J. Illes, B.J. Sahakian, The Oxford Handbook of Neuroethics, Oxford Univ Pr, 2011.

[5] R. Brice, C.L. Teo, Z. Qiang, A. Marcelo, B. Etienne, G. Cuntai, et al., Controlling a wheelchair in a building using thought, IEEE Intell. Syst. (2007) 1–8.

[6] F. Guo, B. Hong, X. Gao, S. Gao, A brain-computer interface using motion-onset visual evoked potential, J. Neural Eng. 5 (2008) 477–485.

[7] R. Leeb, R. Scherer, F. Lee, H. Bischof, G. Pfurtscheller, Navigation in virtual environments through motor imagery, 9th Comput. Vis. Winter Workshop (CVWW) (2004) 99–108.

[8] C. Ware, H.H. Mikaelian, An evaluation of an eye tracker as a device for computer input 2, ACM SIGCHI Bull. 18 (1987) 183–188.

[9] M.R. Williams, R.F. Kirsch, Evaluation of head orientation and neck muscle EMG signals as command inputs to a human–computer interface for individuals with high tetraplegia, IEEE Trans. Neural. Syst. Rehabil. Eng. 16 (2008) 485–496.

[10] G. Bauer, F. Gerstenbrand, E. Rumpl, Varieties of the locked-in syndrome, J. Neurol. 221 (1979) 77–91.

[11] J.R. Patterson, M. Grabois, Locked-in syndrome: a review of 139 cases, Stroke 17 (1986) 758.

[12] A. Kubler, B. Kotchoubey, J. Kaiser, J.R. Wolpaw, N. Birbaumer, Brain-computer communication: unlocking the locked in, Psychol. Bull. 127 (2001) 358–375.

[13] S.G. Mason, G.E. Birch, A general framework for brain-computer interface design, IEEE Trans. Neural. Syst. Rehabil. Eng. 11 (2003) 70–85.

[14] T.M. Vaughan, Guest editorial brain-computer interface technology: a review of the second international meeting, IEEE Trans. Neural. Syst. Rehabil. Eng. 11 (2003) 94–109.

[15] T. Felzer, On the possibility of developing a brain-computer interface (BCI), technical University of Darmstadt, Darmstadt, Germany, Tech. Rep. (2001).

[16] J.R. Wolpaw, N. Birbaumer, W.J. Heetderks, D.J. McFarland, P.H. Peckham, G. Schalk, et al., Brain-computer interface technology: a review of the first international meeting, IEEE Trans. Rehabil. Eng. 8 (2000) 164–173.

[17] G. Dornhege, Toward Brain-Computer Interfacing, The MIT Press, 2007.

[18] J.J. Vidal, Toward direct brain-computer communication, Annu. Rev. Biophys. Bioeng. 2 (1973) 157–180.

[19] J.R. Wolpaw, B. Niels, M. Dennis, G. Pfurtscheller, V. Theresa, Brain-computer interfaces for communication and control, Clin. Neurophysiol. (2002) 767–791.

[20] E.E. Sutter, The brain response interface: communication through visually-induced electrical brain responses, J. Microcomputer Appl. 15 (1992) 31–45.

[21] J.P. Donoghue, Connecting cortex to machines: recent advances in brain interfaces, Nat. Neurosci. 5 (Suppl) (2002) 1085–1088.

[22] F. Galan, M. Nuttin, E. Lew, P.W. Ferrez, G. Vanacker, J. Philips, et al., A brain-actuated wheelchair: asynchronous and non-invasive brain-computer interfaces for continuous control of robots, Clin. Neurophysiol. 119 (2008) 2159–2169.

[23] F. Meng, K. Tong, S. Chan, W. Wong, K. Lui, K. Tang, et al., BCI-FES training system design and implementation for rehabilitation of stroke patients, IEEE Int. Joint Conf. Neural. Netw. (IEEE World Congr. Comput. Intell.) (2008) 4103−4106.

[24] C. Wang, K.S. Phua, K.K. Ang, C. Guan, H. Zhang, R. Lin, et al., A feasibility study of non-invasive motor-imagery BCI-based robotic rehabilitation for stroke patients, in Neural Engineering, 2009. NER'09. 4th International IEEE/EMBS Conference on, (2009) 271−274.

[25] J. Kassubek, A. Unrath, H.J. Huppertz, D. Lulé, T. Ethofer, A.D. Sperfeld, et al., Global brain atrophy and corticospinal tract alterations in ALS, as investigated by voxel-based morphometry of 3-D MRI, Amyotroph. Lat. Scler. 6 (2005) 213−220.

[26] P. Sonksen, S. Hillier, Spinal cord injury, Br. Med. J. 340 (2010) 922−924.

[27] M.G. Fehlings, R.G. Perrin, The timing of surgical intervention in the treatment of spinal cord injury: a systematic review of recent clinical evidence, Spine 31 (2006) S28.

[28] K.S.G. Chua, K.H. Kong, Functional outcome in brain stem stroke patients after rehabilitation, Arch. Phys. Med. Rehabil. 77 (1996) 194−197.

[29] T.K. Daneshmend, Acute brain stem stroke during neck manipulation, Br. Med. J. (Clin. Res. Ed.) 288 (1984) 1090.

[30] H. Aurlien, I.O. Gjerde, J.H. Aarseth, G. Eldoen, B. Karlsen, H. Skeidsvoll, et al., EEG background activity described by a large computerized database, Clin. Neurophysiol. 115 (2004) 665−673.

[31] J.W. Kozelka, T.A. Pedley, Beta and mu rhythms, J. Clin. Neurophysiol. 7 (1990) 191−207.

[32] G. Pfurtscheller F. Lopes da Silva H, Handbook of electroencephalography and clinical neurophysiology − revised series: vol. 6. Event-related desynchronization, Elsevier (1999).

[33] M. Tiefelsdorf, D.A. Griffith, Semiparametric filtering of spatial autocorrelation: the eigenvector approach, Environ. Plann. A 39 (2007) 1193−1221.

[34] B. Blankertz, R. Tomioka, S. Lemm, M. Kawanabe, K.R. Muller, Optimizing spatial filters for robust EEG single-trial analysis, IEEE Signal Process. Mag. 25 (2007) 41−56.

[35] C.S.L. Tsui, J.Q. Gan, S.J. Roberts, A self-paced brain-computer interface for controlling a robot simulator: an online event labelling paradigm and an extended Kalman filter based algorithm for online training, Med. Biol. Eng. Comput. 47 (2009) 257−265.

[36] V. Gandhi, V. Arora, L. Behera, G. Prasad, D. Coyle, T.M. McGinnity, A recurrent quantum neural network model enhances the EEG signal for an improved brain-computer interface, Assist. Living, Inst. Eng. Technol. Conf. (2011).

[37] V. Gandhi, V. Arora, G. Prasad, D. Coyle, T.M. McGinnity, A novel EEG signal enhancement approach using a recurrent quantum neural network for a brain-computer interface, 3rd European Conference, Technically Assisted Rehabilitation, Berlin, 2011.

[38] V. Jeyabalan, A. Samraj, L.C. Kiong, Motor imaginary signal classification using adaptive recursive bandpass filter and adaptive autoregressive models for brain machine interface designs, Int. J. Biol. Med. Sci. 3 (2008) 236−243.

[39] D. Coyle, Intelligent Preprocessing and Feature Extraction Techniques for a Brain Computer Interface, PhD Thesis, University of Ulster, 2006.

[40] T. Lan, D. Erdogmus, A. Adami, S. Mathan, M. Pavel, Channel selection and feature projection for cognitive load estimation using ambulatory EEG, Comput. Intell. Neurosci. 2007 (2007) 8.

[41] S.V. Ramanan, N. Kalpakam, J. Sahambi, A novel wavelet based technique for detection and de-noising of ocular artifact in normal and epileptic electroencephalogram, pp. 1027−1031. International Conference on Communications, Circuits and Systems, 2004.

[42] C. Yamaguchi, Fourier and wavelet analyses of normal and epileptic electroencephalogram (EEG), First Int. IEEE EMBS Conf. Neural Eng. (2003) 406–409.

[43] R.J. Croft, R.J. Barry, Removal of ocular artifact from the EEG: a review, Neurophysiol. Clin. 30 (2000) 5–19.

[44] P. Shenoy, M. Krauledat, B. Blankertz, R.P. Rao, K.R. Muller, Towards adaptive classification for BCI, J. Neural. Eng. 3 (2006) R13–23.

[45] J. Kronegg, G. Chanel, S. Voloshynovskiy, T. Pun, EEG-based synchronized brain-computer interfaces: a model for optimizing the number of mental tasks, IEEE Trans. Neural Syst. Rehabil. Eng. 15 (2007) 50–58.

[46] D. Coyle, G. Prasad, T.M. McGinnity, A time-series prediction approach for feature extraction in a brain-computer interface, IEEE Trans. Neural. Syst. Rehabil. Eng. 13 (2005) 461–467.

[47] D. Coyle, G. Prasad, T.M. McGinnity, Extracting features for a brain-computer interface by self-organising fuzzy neural network-based time series prediction, 26th Annu. Int. Conf. IEEE Eng. Med. Biol. Soc. (2004).

[48] D. Coyle, G. Prasad, T.M. McGinnity, A time-frequency approach to feature extraction for a brain-computer interface with a comparative analysis of performance measures, EURASIP J. Appl. Signal Processing 19 (2005) 3141–3151.

[49] P. Herman, G. Prasad, T.M. McGinnity, D. Coyle, Comparative analysis of spectral approaches to feature extraction for EEG-based motor imagery classification, IEEE Trans. Neural. Syst. Rehabil. Eng. 16 (2008) 317–326.

[50] J. Pari, K. Vassilis, R. Kenneth, EEG signal classification using wavelet feature extraction and neural networks, IEEE John Vincent Atanasoff 2006 Int. Symp. Mod. Comput. (2006).

[51] D.J. McFarland, C.W. Anderson, K.R. Muller, A. Schlogl, D.J. Krusienski, BCI meeting 2005 – workshop on BCI signal processing: feature extraction and translation, IEEE Trans. Neural Syst. Rehabil. Eng. 14 (2006) 135–138.

[52] S. Cososchi, R. Strungaru, A. Ungureanu, M. Ungureanu, EEG feature extraction for motor imagery, 28th Annu. Int. Conf. IEEE Eng. Med. Biol. Soc. 2006 (2008) 1142–1145. EMBS'06.

[53] B. Hjorth, EEG analysis based on time domain properties, Electroencephalogr. Clin. Neurophysiol. 29 (1970) 306–310.

[54] I.S. Rao, V.K. Anandan, Bispectral analysis of atmospheric radar signals, IEEE Aerosp. Electron. Syst. Mag. 23 (2008) 38–41.

[55] V. Gandhi, D. Coyle, G. Prasad, C. Bharti, L. Behera T. McGinnity M., Interfacing a dynamic interface paradigm for multiple target selection using a two class brain computer interface, in Indo - US Workshop on System of Systems Engineering, (2009).

[56] V. Gandhi, G. Prasad, D. Coyle, L. Behera, T.M. McGinnity, A novel paradigm for multiple target selection using a two class brain computer interface, IET Ir. Signals Syst. Conf. (ISSC 2009) (2010) 1–6.

[57] V. Gandhi, G. Prasad, D. Coyle, L. Behera, T.M. McGinnity, An intelligent adaptive user interface (iAUI) for enhancing the communication in a brain-computer interface (BCI), Int. UKIERI Workshop on Fusion Brain-Comput. Interface Assist. Rob. (2011) 26.

[58] J. D. Bayliss, A flexible brain-computer interface, Ph.D. Thesis, University of Rochester, 2001.

[59] M.F. Bear, B.W. Connors, M.A. Paradiso, Neuroscience: Exploring the Brain, Lippincott Williams & Wilkins, 2007.

[60] R. Krepkiy, Brain-Computer Interfaces: Design and Implementation of an Online BCI System for the Control in Gaming Applications an Virtual Limbs. VDM Verlag Dr. Muller, 2008.

[61] J. Decety, The neurophysiological basis of motor imagery, Behav. Brain Res. 77 (1996) 45–52.

[62] M. Jeannerod, M.A. Arbib, G. Rizzolatti, H. Sakata, Grasping objects: the cortical mechanisms of visuomotor transformation, Trends Neurosci. 18 (1995) 314–320.

[63] g. Tec EEG and cap [Online]. Available: <http://www.gtec.at/Products/Electrodes-and-Sensors/g.SAHARA-Specs-Features> (accessed 16.07.14).

[64] E. Niedermeyer, F.H.L. Da Silva, Electroencephalography: Basic Principles, Clinical Applications, and Related Fields, Lippincott Williams & Wilkins, 2004.

[65] B. He, XVI, 488 p.203 Illus. with CD-ROM. 0–306–48609–1 in: B. He (Ed.), Neural Engineering, Neural Engineering, Springer, Berlin, 2005.

[66] K. Blinowska, P. Durka, Electroencephalography (EEG), Wiley Encycl. Biomed. Eng. (2006).

[67] M. Teplan, Fundamentals of EEG measurement, Meas. Sci. Rev. 2 (2002) 1–11.

[68] M.A. Kisley, Z.M. Cornwell, Gamma and beta neural activity evoked during a sensory gating paradigm: effects of auditory, somatosensory and cross-modal stimulation, Clin. Neurophysiol. 117 (2006) 2549–2563.

[69] M. Grosse-Wentrup, B. Schölkopf, J. Hill, User research causal influence of gamma oscillations on the sensorimotor-rhythm, Brain 35 (2010).

[70] M. Grosse-Wentrup, B. Schölkopf, J. Hill, Causal influence of gamma oscillations on the sensorimotor rhythm, Neuroimage 56 (2011) 837–842.

[71] SG Dastidar, Models of EEG Data Mining and Classification in Temporal Lobe Epilepsy: Wavelet-Chaos-Neural Network Methodology and Spiking Neural Networks, PhD. Thesis, The Ohio State University, 2007.

[72] C. Tsui, J. Gan, Asynchronous BCI control of a robot simulator with supervised online training, Intell. Data Eng. Autom. Learn. -IDEAL 2007 (2007) 125–134.

[73] G. Krausz, R. Scherer, G. Korisek, G. Pfurtscheller, Critical decision-speed and information transfer in the Graz Brain-computer Interface, Appl. Psychophysiol. Biofeedback 28 (2003) 233–240.

[74] d.e.l. José, R. Millán, Adaptive brain interfaces, Commun. ACM 46 (2003).

[75] D. Choi, Y. Ryu, Y. Lee, M. Lee, D. Choi, Y. Ryu, et al., Performance evaluation of a motor-imagery-based EEG-Brain computer interface using a combined cue with heterogeneous training data in BCI-Naive subjects, Biomed. Eng. Online 10 (2011) 1–12.

[76] G. Townsend, B. Graimann, G. Pfurtscheller, Continuous EEG classification during motor imagery-simulation of an asynchronous BCI, Neural Systems and Rehabilitation Engineering, IEEE Transactions on, 12, (2004) 258–265.

[77] R. Leeb, H. Sagha, R. Chavarriaga, J.R. Millán, A hybrid brain-computer interface based on the fusion of electroencephalographic and electromyographic activities, J. Neural Eng. 8 (2011) 025011.

[78] T. Solis-Escalante, G. Müller-Putz, C. Brunner, V. Kaiser, G. Pfurtscheller, Analysis of sensorimotor rhythms for the implementation of a brain switch for healthy subjects, Biomed. Signal Process. Control 5 (2010) 15–20.

[79] M. Krauledat, B. Blankertz, G. Dornhege, M. Schroeder, G. Curio, K.R. Müller, On-line differentiation of neuroelectric activities: Algorithms and applications, 28th Annu. Int. Conf. IEEE Eng. Med. Biol. Soc. (2006).

[80] T. Geng, J.Q. Gan, Motor prediction in brain-computer interfaces for controlling mobile robots, 30th Annu. Int. Conf. IEEE Eng. Med. Biol. Soc. (2008) 634–637.

[81] E.B. Sadeghian, M.H. Moradi, Continuous detection of motor imagery in a four-class asynchronous BCI, 29th Annu. Int. Conf. IEEE Eng. Med. Biol. Soc. 2007 (2007) 3241–3244.

[82] A. Satti, D. Coyle, G. Prasad, Continuous EEG classification for a self-paced BCI, 4th Int. IEEE/EMBS Conf. Neural Eng. (2009) 315–318.

[83] M. Roth, J. Decety, M. Raybaudi, R. Massarelli, C. Delon-Martin, C. Segebarth, et al., Possible involvement of primary motor cortex in mentally simulated movement: a functional magnetic resonance imaging study, Neuroreport 7 (1996) 1280.

[84] J. Decety, M. Jeannerod, D. Durozard, G. Baverel, Central activation of autonomic effectors during mental simulation of motor actions in man, J. Physiol. (Lond.) 461 (1993) 549.

[85] G. Yue, K.J. Cole, Strength increases from the motor program: comparison of training with maximal voluntary and imagined muscle contractions, J. Neurophysiol. 67 (1992) 1114.

[86] S. Silbernagl, F. Lang, Color Atlas of Pathophysiology, Thieme Medical Pub, 2009.

[87] J. Malmivuo, R. Plonsey, Bioelectromagnetism: Principles and Applications of Bioelectric and Biomagnetic Fields, Oxford University Press, USA, 1995.

[88] H. Jasper, The ten-twenty electrode system of the International federation, Electroencephalogr. Clin. Neurophysiol. 10 (1958) 371–375.

[89] F. Sharbrough, G. Chatrian, R. Lesser, H. Lüders, M. Nuwer, T. Picton, American Electroencephalographic Society guidelines for standard electrode position nomenclature, J. Clin. Neurophysiol. 8 (1991) 200–202.

[90] R.T. Pivik, R.J. Broughton, R. Coppola, R.J. Davidson, N. Fox, M.R. Nuwer, Guidelines for the recording and quantitative analysis of electroencephalographic activity in research contexts, Psychophysiology 30 (1993) 547–558.

[91] J.T. Cacioppo, L.G. Tassinary, G.G. Berntson, Handbook of Psychophysiology, Cambridge Univ Pr, 2007.

[92] J.T. Farrow, J.R. Hebert, Breath suspension during the Transcendental Meditation technique, Psychosom. Med. 44 (1982) 133–153.

[93] Neurodiagnostic policies and procedures manual. 2010. [Online]. Available: <http://neurology.stanford.edu/PDFs/neurodiagnostics_policies_manual_2010.pdf> (accessed April 2014). Also available on: (accessed 16.07.14) <http://www.scribd.com/doc/152597824/Stanford-Neurodiagnostics-Policies-Manual-2010>.

[94] Electroencephalography [Online]. Available: <http://en.wikipedia.org/wiki/Electroencephalography> (accessed 16.07.14).

[95] E. Wyllie, A. Gupta, D.K. Lachhwani, The Treatment of Epilepsy: Principles and Practice, Lippincott Williams & Wilkins, USA, 2005.

[96] C. Guger, Brain Computer Interface with g.MOBIlab and Simulink, g.Tec. Available online at: <http://www.gtec.at/content/download/1859/11556/version/4/#> (accessed 16.07.14).

[97] A. Bashashati, M. Fatourechi, R.K. Ward, G.E. Birch, A survey of signal processing algorithms in brain-computer interfaces based on electrical brain signals, J. Neural Eng. 4 (2007) R32–57.

[98] S. Sutton, M. Braren, J. Zubin, E.R. John, Evoked-potential correlates of stimulus uncertainty, Science 150 (1965) 1187.

[99] L.A. Farwell, E. Donchin, Talking off the top of your head: toward a mental prosthesis utilizing event-related brain potentials, Electroencephalogr. Clin. Neurophysiol. 70 (1988) 510–523.

[100] E. Donchin, K.M. Spencer, R. Wijesinghe, The mental prosthesis: assessing the speed of a P300-based brain-computer interface, IEEE Trans. Rehabil. Eng. 8 (2000) 174–179.

[101] A. Nijholt, D. Tan, Brain-computer interfacing for intelligent systems, IEEE Intell. Syst. 23 (2008) 72–79.

[102] P.W. Ferrez, J. del, R Millan, Error-related EEG potentials generated during simulated brain-computer interaction, IEEE Trans. Biomed. Eng. 55 (2008) 923–929.

[103] G.R. Muller-Putz, R. Scherer, C. Neuper, G. Pfurtscheller, Steady-state somatosensory evoked potentials: suitable brain signals for brain-computer interfaces? IEEE Trans. Neural Syst. Rehabil. Eng. 14 (2006) 30–37.

[104] J.J. Vidal, Real-time detection of brain events in EEG, Proc. IEEE 65 (1977) 633–641.

[105] S.T. Morgan, J.C. Hansen, S.A. Hillyard, Selective attention to stimulus location modulates the steady-state visual evoked potential, Proc. Natl. Acad. Sci. 93 (1996) 4770.

[106] M.M. Müller, T.W. Picton, P. Valdes-Sosa, J. Riera, W.A. Teder-Sälejärvi, S.A. Hillyard, Effects of spatial selective attention on the steady-state visual evoked potential in the 20−28 Hz range, Cogn. Brain Res. 6 (1998) 249−261.

[107] M. Kuba, Z. Kubova, J. Kremlacek, J. Langrova, Motion-onset VEPs: characteristics, methods, and diagnostic use, Vision Res. 47 (2007) 189−202.

[108] S.P. Heinrich, A primer on motion visual evoked potentials, Doc. Ophthalmol. 114 (2007) 83−105.

[109] J. Kremlcek, M. Hulan, M. Kuba, Z. Kubov, J. Langrov, F. Vt, et al., Role of latency jittering correction in motion-onset VEP amplitude decay during prolonged visual stimulation, Doc. Ophthalmol. (2012) 1−13.

[110] B. Hong, F. Guo, T. Liu, X. Gao, S. Gao, N200-speller using motion-onset visual response, Clin. Neurophysiol. 120 (2009) 1658−1666.

[111] T. Hinterberger, J.M. Houtkooper, B. Kotchoubey, Effects of feedback control on slow cortical potentials and random events, Parapsychol. Assoc. Conv. (2004) 39−50.

[112] N. Birbaumer, T. Elbert, A.G. Canavan, B. Rockstroh, Slow potentials of the cerebral cortex and behavior, Physiol. Rev. 70 (1990) 1.

[113] N. Neumann, J. Kaiser, B. Kotchoubey, T. Hinterberger, N.P. Birbaumer, Brain-computer communication: self-regulation of slow cortical potentials for verbal communication, Arch. Phys. Med. Rehabil. 82 (2001) 1533−1539.

[114] F. Nijboer, J. Mellinger, T. Matuz, U. Mochty, E. Sellers, T.M. Vaughan, et al., Comparing sensorimotor rhythms, slow cortical potentials, and P300 for brain-computer interface (BCI) use by ALS patients, Presented at 3rd Int. BCI Conf. (2005).

[115] T. Hinterberger, S. Schmidt, N. Neumann, J. Mellinger, B. Blankertz, G. Curio, et al, Brain-computer communication and slow cortical potentials, Biomedical Engineering, IEEE Transactions on, 51, (2004) 1011−1018.

[116] G. Pfurtscheller, F.H. Lopes da Silva, Event-related EEG/MEG synchronization and desynchronization: basic principles, Clin. Neurophysiol. 110 (1999) 1842−1857.

[117] G. Pfurtscheller, A. Aranibar, Event-related cortical desynchronization detected by power measurements of scalp EEG, Electroencephalogr. Clin. Neurophysiol. 42 (1977) 817−826.

[118] G. Pfurtscheller, Graphical display and statistical evaluation of event-related desynchronization (ERD), Electroencephalogr. Clin. Neurophysiol. 43 (1977) 757−760.

[119] G. Pfurtscheller, Event-related synchronization (ERS): an electrophysiological correlate of cortical areas at rest, Electroencephalogr. Clin. Neurophysiol. vol. 83 (1992) 62−69.

[120] J.R. Wolpaw, D.J. McFarland, Multichannel EEG-based brain-computer communication, Electroencephalogr. Clin. Neurophysiol. 90 (1994) 444−449.

[121] G. Pfurtscheller, C. Neuper, Motor imagery and direct brain-computer communication, Proc. IEEE 89 (2001) 1123−1134.

[122] G. Pfurtscheller, C. Neuper, D. Flotzinger, M. Pregenzer, EEG-based discrimination between imagination of right and left hand movement, Electroencephalogr. Clin. Neurophysiol. 103 (1997) 642−651.

[123] D.J. McFarland, J.R. Wolpaw, Sensorimotor rhythm-based brain-computer interface (BCI): model order selection for autoregressive spectral analysis, J. Neural Eng. 5 (2008) 155.

[124] D.J. McFarland, J.R. Wolpaw, Sensorimotor rhythm-based brain-computer interface (BCI): feature selection by regression improves performance, Neural Systems and Rehabilitation Engineering, IEEE Transactions 13 (2005) 372−379.

[125] D.J. McFarland, L.A. Miner, T.M. Vaughan, J.R. Wolpaw, Mu and beta rhythm topographies during motor imagery and actual movements, Brain Topogr. 12 (2000) 177−186.

[126] F. Pichiorri, F.D.V. Fallani, F. Cincotti, F. Babiloni, M. Molinari, S. Kleih, et al., Sensorimotor rhythm-based brain-computer interface training: the impact on motor cortical responsiveness, J. Neural Eng. 8 (2011) 025020.

[127] A. Kübler, F. Nijboer, J. Mellinger, T.M. Vaughan, H. Pawelzik, G. Schalk, et al., Patients with ALS can use sensorimotor rhythms to operate a brain-computer interface, Neurology 64 (2005) 1775–1777.

[128] E.V.C. Friedrich, D.J. McFarland, C. Neuper, T.M. Vaughan, P. Brunner, J.R. Wolpaw, A scanning protocol for a sensorimotor rhythm-based brain-computer interface, Biol. Psychol. 80 (2009) 169–175.

[129] A.S. Royer, A. McCullough, B. He, A sensorimotor rhythm based goal selection brain-computer interface, in Engineering in Medicine and Biology Society, 2009. EMBC 2009, Annu. Int. Conf. IEEE (2009) 575–577.

[130] D.J. Krusienski, D.J. McFarland, J.R. Wolpaw, Value of amplitude, phase, and coherence features for a sensorimotor rhythm-based brain-computer interface, Brain Res. Bull. 87 (2012) 130–134.

[131] S. Halder, D. Agorastos, R. Veit, E. Hammer, S. Lee, B. Varkuti, et al., Neural mechanisms of brain-computer interface control, Neuroimage (2011).

[132] G. Pfurtscheller, C. Neuper, C. Guger, W. Harkam, H. Ramoser, A. Schlogl, et al., Current trends in Graz brain-computer interface (BCI) research, IEEE Trans. Rehabil. Eng. 8 (2000) 216–219.

[133] N. Birbaumer, N. Ghanayim, T. Hinterberger, I. Iversen, B. Kotchoubey, A. Kuebler, et al., A spelling device for the paralysed, Nature 398 (1999) 297–298.

[134] G.R. McMillan, G.L. Calhoun, M.S. Middendorf, J.H. Schnurer, D.F. Ingle, V.T. Nasman, Direct brain interface utilizing self-regulation of steady-state visual evoked response (SSVER), Proc. RESNA Conf. 1995 (1995) 693–695, Washington.

[135] M. Middendorf, G. McMillan, G. Calhoun, K.S. Jones, Brain-computer interfaces based on the steady-state visual-evoked response, IEEE Trans. Rehabil. Eng. 8 (2000) 211–214.

[136] A. Delorme, T. Sejnowski, S. Makeig, Enhanced detection of artifacts in EEG data using higher-order statistics and independent component analysis, Neuroimage 34 (2007) 1443–1449.

[137] S. Romero, M.A. Mañanas, M.J. Barbanoj, Ocular reduction in EEG signals based on adaptive filtering, regression and blind source separation, Ann. Biomed. Eng. 37 (2009) 176–191.

[138] A.Y.K. Chan, Biomedical Device Technology: Principles and Design, Charles C Thomas Pub Ltd, 2008.

[139] M.E. Calkins, W.G. Iacono, C.E. Curtis, Smooth pursuit and antisaccade performance evidence trait stability in schizophrenia patients and their relatives, Int. J. Psychophysiol. 49 (2003) 139–146.

[140] T.P. Jung, C. Humphries, T.W. Lee, S. Makeig, M.J. McKeown, V. Iragui, et al., Extended ICA removes artifacts from electroencephalographic recordings, Adv. Neural Inf. Process. Syst. (1998) 894–900.

[141] C. Guger, H. Ramoser, G. Pfurtscheller, Real-time EEG analysis with subject-specific spatial patterns for a brain-computer interface (BCI), IEEE Trans. Rehabil. Eng. 8 (2000) 447–456.

[142] J. Qin, Y. Li, A. Cichocki, ICA and committee machine-based algorithm for cursor control in a BCI system, Adv. Neural Netw. – ISNN 2005 (2005) 293–318.

[143] D. Gutiérrez, Using single/multi-channel energy transform as preprocessing tool for magnetoencephalographic data-based applications, pp. 114–118 20th International Conference on Electronics, Communications and Computer (CONIELECOMP), 2010

[144] P.L. Nunez, R.B. Silberstein, P.J. Cadusch, R.S. Wijesinghe, A.F. Westdorp, R. Srinivasan, A theoretical and experimental study of high resolution EEG based on

surface Laplacians and cortical imaging, Electroencephalogr. Clin. Neurophysiol. 90 (1994) 40−57.

[145] D.J. McFarland, L.M. McCane, S.V. David, J.R. Wolpaw, Spatial filter selection for EEG-based communication, Electroencephalogr. Clin. Neurophysiol. 103 (1997) 386−394.

[146] K. Pearson LIII, On lines and planes of closest fit to systems of points in space, Philos. Mag. Ser. 6 2 (1901) 559−572.

[147] H. Lee, S. Choi, PCA + HMM + SVM for eeg pattern classification, Proc. 7th Int. Symp. Signal Process. Appl. 1 (2003) 541−544.

[148] C. Jutten, J. Herault, Blind separation of sources, part I: An adaptive algorithm based on neuromimetic architecture, Signal Process. 24 (1991) 1−10.

[149] A. Hyvärinen, J. Karhunen, E. Oja, Independent Component Analysis, Wiley-Interscience, 2001.

[150] P. Comon, Independent component analysis, a new concept? Signal Process. 36 (1994) 287−314.

[151] S. Makeig, A.J. Bell, T.P. Jung, T.J. Sejnowski, Independent component analysis of electroencephalographic data, Adv. Neural Inf. Process. Syst. (1996) 145−151.

[152] A. Hyvarinen, P. Pajunen, On existence and uniqueness of solutions in nonlinear independent component analysis, Proc. IEEE Int. Joint Conf. Neural Netw., IEEE World Congr. Comput. Intell. 2 (1998) 1350−1355.

[153] P. Pajunen, J. Karhunen, A maximum likelihood approach to nonlinear blind source separation, Artif. Neural Netw. − ICANN'97 (1997) 541−546.

[154] S. Wang, Enhancing brain-computer interfacing through advanced independent component analysis techniques. Ph. D. Thesis, University of Southampton, 2009.

[155] T.P. Jung, S. Makeig, C. Humphries, T.W. Lee, M.J. Mckeown, V. Iragui, et al., Removing electroencephalographic artifacts by blind source separation, Psychophysiology 37 (2000) 163−178.

[156] W. Ting, Y. Guo-zheng, Y. Bang-hua, S. Hong, EEG feature extraction based on wavelet packet decomposition for brain computer interface, Measurement 41 (2008) 618−625.

[157] A. Malatesta, L.R. Quitadamo, M. Abbafati, L. Bianchi, M.G. Marciani, G.C. Cardarilli, Moving towards a hardware implementation of the independent component analysis for brain computer interfaces, IEEE Conf. Biomed. Circuits Syst. Conf. (2007) 227−230.

[158] Z.J. Koles, The quantitative extraction and topographic mapping of the abnormal components in the clinical EEG, Electroencephalogr. Clin. Neurophysiol. 79 (1991) 440−447.

[159] J. Müller-Gerking, G. Pfurtscheller, H. Flyvbjerg, Designing optimal spatial filters for single-trial EEG classification in a movement task, Clin. Neurophysiol. 110 (1999) 787−798.

[160] H. Ramoser, J. Muller-Gerking, G. Pfurtscheller, Optimal spatial filtering of single trial EEG during imagined hand movement, IEEE Trans. Rehabil. Eng. vol. 8 (2000) 441−446.

[161] T. Al-ani and D. Trad, Signal processing and classification approaches for brain-computer interface, Intelligent and Biosensors, Edited by Vernon S. Somerset, pp. 25−66, 2010. Available from: <http://www.intechopen.com/books/intelligent-and-biosensors/signal-processing-and-classification-approaches-for-brain-computer-interface> (accessed 16.07.14).

[162] K.K. Ang, Z.Y. Chin, H. Zhang, C. Guan, Filter bank common spatial pattern (FBCSP) in brain-computer interface, IEEE Int. Joint Conf. Neural Networks (IEEE World Congress on Comput. Intell.) (2008) 2390−2397.

[163] S. Lemm, B. Blankertz, G. Curio, K.R. Muller, Spatio-spectral filters for improving the classification of single trial EEG, IEEE Trans. Biomed. Eng. 52 (2005) 1541−1548.

[164] F. Lotte, C. Guan, Regularizing common spatial patterns to improve BCI designs: unified theory and new algorithms, IEEE Trans. Biomed. Eng. (2011) 1.

[165] G. Dornhege, B. Blankertz, M. Krauledat, F. Losch, G. Curio, K.R. Muller, Combined optimization of spatial and temporal filters for improving brain-computer interfacing, IEEE Trans. on Biomed. Eng. 53 (2006) 2274–2281.

[166] G. Dornhege, B. Blankertz, M. Krauledat, F. Losch, G. Curio, K. Muller, Optimizing spatio-temporal filters for improving brain-computer interfacing, Adv. Neural Inf. Process. Syst. 18 (2006) 315.

[167] Q. Novi, C. Guan, T.H. Dat, P. Xue, Sub-band common spatial pattern (SBCSP) for brain-computer interface, 3rd Int. IEEE/EMBS Conf. Neural Eng. - CNE'07 (2007) 204–207.

[168] A. Satti, D. Coyle, G. Prasad, Optimal frequency band selection with particle swarm optimization for a brain computer interface, Proceedings of the Workshop on Evolutionary Computing Lecture Series by Pioneers, University of Ulster, 2008.

[169] D. Coyle, A. Satti, G. Prasad, T.M. McGinnity, Neural time-series prediction preprocessing meets common spatial patterns in a brain-computer interface, 30th Annu. Int. Conf. IEEE Eng. Med. Biol. Soc. (2008) 2626–2629.

[170] D. Coyle, T.M. McGinnity, G. Prasad, Creating a nonparametric brain-computer interface with neural time-series prediction preprocessing, 28th Annu. Int. Conf. IEEE Eng. Med. Biol. Soc. (2006) 2183–2186.

[171] D. Coyle, Recent advances in prediction-based eeg preprocessing for improved brain-computer interface performance, vol. Intech, New Developments in Biomedical Engineering (2010) 123–150.

[172] D. Coyle, Neural network based auto association and time-series prediction for bio-signal processing in brain-computer interfaces, IEEE Comput. Intell. Mag. 4 (2009) 47–59.

[173] R.E. Kalman, A new approach to linear filtering and prediction problems, J. Basic Eng. 82 (1960) 35–45.

[174] L. Kleeman, Understanding and applying Kalman filtering, 25–26 January 1996 Proceedings of the Second Workshop on Perceptive Systems, Curtin University of Technology, Perth Western Australia, 1996.

[175] G. Welch, G. Bishop, An introduction to the Kalman Filter, vol. 7, University of North Carolina at Chapel Hill, Chapel Hill, NC, 1995.

[176] W. Malik, W. Truccolo, E. Brown, L. Hochberg, Efficient decoding with steady-state Kalman filter in neural interface systems, IEEE Trans. Neural Syst. Rehabil. Eng. (2011) 1.

[177] M. Kamrunnahar, S. Schiff, A square root ensemble Kalman filter application to a motor-imagery brain-computer interface, 2011 Annu. Int. Conf. IEEE Eng. Med. Biol. Soc. (2011) 6385–6388.

[178] S.S. Haykin, Kalman Filtering and Neural Networks, Wiley Online Library, 2001.

[179] S.J. Julier, J.K. Uhlmann, A new extension of the Kalman filter to nonlinear systems, Int. Symp. Aerospace/Defense Sensing, Simul. Controls 26 (1997).

[180] G. Prasad, P. Herman, D. Coyle, C. Jacqueline, Applying a brain-computer interface to support motor imagery practice in people with stroke for upper limb recovery: a feasibility study, J. NeuroEng. Rehabil. 7 (2010) 60.

[181] M. Deriche, AR parameter estimation from noisy data using the EM algorithm, IEEE International Conference on Acoustics, Speech, and Signal Processing, pp. IV/69-IV/72 vol. 4, 1994.

[182] Y. Xia, M.S. Kamel, H. Leung, A fast algorithm for AR parameter estimation using a novel noise-constrained least-squares method, Neural Networks 23 (2010) 396–405.

[183] S.M. Kay, S.L. Marple Jr, Spectrum analysis – a modern perspective, Proc. IEEE 69 (1981) 1380–1419.

[184] A. Schlogl, F. Lee, H. Bischof, G. Pfurtscheller, Characterization of four-class motor imagery EEG data for the BCI-competition, J. Neural Eng. 2 (2005) L14–22.

[185] R.O. Duda, P.E. Hart, D.G. Stork, Pattern classification, Citeseer (2001).

[186] M. Vourkas, S. Micheloyannis, G. Papadourakis, Use of ANN and Hjorth parameters in mental task discrimination, First Int. Conf. Adv. Med. Signal Inf. Process. (2000) 327–332.

[187] C. Vidaurre, A. Schlogl, R. Cabeza, R. Scherer, G. Pfurtscheller, Study of on-line adaptive discriminant analysis for EEG-based brain computer interfaces, IEEE Trans. Biomed. Eng. 54 (2007) 550–556.

[188] R. Boostani, M.H. Moradi, A new approach in the BCI research based on fractal dimension as feature and Adaboost as classifier, J. Neural Eng. 1 (2004) 212–217.

[189] F. Boller, J. Grafman, Handbook of Neuropsychology, Elsevier Science Health Science Div, 2000.

[190] P. Stoica, R.L. Moses, Introduction to Spectral Analysis, Prentice Hall, Upper Saddle River, NJ, 1997.

[191] F. Babiloni, F. Cincotti, L. Lazzarini, J. Millan, J. Mourino, M. Varsta, et al., Linear classification of low-resolution EEG patterns produced by imagined hand movements, IEEE Trans. Rehabil. Eng. 8 (2000) 186–188.

[192] E. Yom-Tov, G.F. Inbar, Feature selection for the classification of movements from single movement-related potentials, IEEE Trans. Neural Syst. Rehabil. Eng. 10 (2002) 170–177.

[193] C. Vidaurre, A. Schlögl, R. Cabeza, R. Scherer, G. Pfurtscheller, Adaptive on-line classification for EEG-based brain computer interfaces with AAR parameters and band power estimates/adaptive on-line classification, einer EEG-basierenden Gehirn-Computer Schnittstelle mit Adaptive Autoregressiven und Bandleistungsparametern, Biomed. Tech. (Berl.) 50 (2005) 350–354.

[194] A.P. Liavas, G.V. Moustakides, G. Henning, E.Z. Psarakis, P. Husar, A periodogram-based method for the detection of steady-state visually evoked potentials, IEEE Trans. Biomed. Eng. 45 (1998) 242–248.

[195] P. Herman, Computational Intelligence Approaches to Handling Uncertainty in the Analysis of Brain Signals, Ph. D. Thesis, University of Ulster, 2008.

[196] A. Bashashati, R.K. Ward, G.E. Birch, Towards development of a 3-state self-paced brain-computer interface, Comput. Intell. Neurosci. 2007 (2007), Article ID 84386, Available from: http://dx.doi.org/10.1155/2007/84386.

[197] J.D. Bronzino, W.A. Hyman, Biomedical signal analysis, in The Biomedical Engineering Handbook, Pergamon Press, New York, 1996.

[198] M. Akay, C. Mello, Wavelets for biomedical signal processing, Proc. 19th Annu. Int. Conf. IEEE Eng. Med. Biol. Soc. 6 (1997) 2688–2691.

[199] S. Blanco, R.Q. Quiroga, O. Rosso, S. Kochen, Time-frequency analysis of electroencephalogram series, Phys. Rev. E 51 (1995) 2624.

[200] S. G. Mallat, A theory for multiresolution signal decomposition: The wavelet representation, Pattern Analysis and Machine Intelligence, IEEE Transactions 11 (1989) 674–693.

[201] A.A. Petrosian, R. Homan, D. Prokhorov, D.C. Wunsch II, Classification of epileptic EEG using neural network and wavelet transform, Proc. SPIE (1996) 834.

[202] A.A. Petrosian, D. Prokhorov, R.B. Schiffer, Early recognition of Alzheimer's disease in EEG using recurrent neural network and wavelet transform, Proc. SPIE (2000) 870.

[203] H. Adeli, S. Ghosh-Dastidar, N. Dadmehr, A wavelet-chaos methodology for analysis of EEGs and EEG subbands to detect seizure and epilepsy, Biomedical Engineering, IEEE Transactions 54 (2007) 205–211.

[204] H. Adeli, Z. Zhou, N. Dadmehr, Analysis of EEG records in an epileptic patient using wavelet transform, J. Neurosci. Methods 123 (2003) 69–87.

[205] S. Ghosh-Dastidar, H. Adeli and N. Dadmehr, Mixed-band wavelet-chaos-neural network methodology for epilepsy and epileptic seizure detection, Biomedical Engineering, IEEE Transactions 54 (2007) 1545–1551.

[206] Z. Zhou, B. Wan, Wavelet packet-based independent component analysis for feature extraction from motor imagery EEG of complex movements, Clin. Neurophys. (2012).

[207] H. Işik, E. Sezer, Diagnosis of epilepsy from electroencephalography signals using multilayer perceptron and Elman artificial neural networks and wavelet transform, J. Med. Syst. 36 (2012) 1.

[208] I. Daubechies, The wavelet transform, time-frequency localization and signal analysis, IEEE Transactions Inf. Theory 36 (1990) 961–1005.

[209] D. Cvetkovic, E.D. Übeyli, I. Cosic, Wavelet transform feature extraction from human PPG, ECG, and EEG signal responses to ELF PEMF exposures: a pilot study, Digital Signal Process. 18 (2008) 861–874.

[210] U. Hoffmann, Bayesian Machine Learning Applied in a Brain-Computer Interface for Disabled Users, Ph.D. dissertation, Ecole Polytechnique Federale de Lausanne, Switzerland, 2007.

[211] D. Coyle, G. Prasad, T.M. McGinnity, Faster self-organizing fuzzy neural network training and a hyperparameter analysis for a brain-computer interface, IEEE Transactions on systems, man, and cybernetics, Part B: Cybern. 39 (2009) 1458–1471.

[212] C. Vidaurre, N. Kramer, B. Blankertz, A. Schlogl, Time domain parameters as a feature for EEG-based brain-computer interfaces, Neural Networks 22 (2009) 1313–1319.

[213] V. Gandhi, V. Arora, L. Behera, G. Prasad, D. Coyle, T. McGinnity, EEG denoising with a recurrent quantum neural network for a brain-computer interface, The 2011 Int. Joint Conf. Neural Networks (IJCNN) (2011) 1583–1590.

[214] C. Guger, A. Schlogl, C. Neuper, D. Walterspacher, T. Strein, G. Pfurtscheller, Rapid prototyping of an EEG-based brain-computer interface (BCI), IEEE Trans, Neural Syst. Rehabil. Eng. 9 (2001) 49–58.

[215] N.E. Huang, Z. Shen, S.R. Long, M.C. Wu, H.H. Shih, Q. Zheng, et al., The empirical mode decomposition and the Hilbert spectrum for nonlinear and non-stationary time series analysis, Proc. R. Soc. Lond. A Math. Phys. Sci. 454 (1998) 903–995.

[216] N.E. Huang, Introduction to the Hilbert Huang Transform, World Scientific Publishing Co. Pte. Ltd., 2005.

[217] N.E. Huang, Introduction to the Hilbert Huang Transform and its related mathematical problems, Hilbert-Huang Transform Appl. 5 (2005) 1–26.

[218] N.E. Huang, S.S. Shen, Hilbert-Huang Transform and its Applications, World Scientific Pub Co Inc, 2005.

[219] N. Brodu, F. Lotte and A. Lécuyer, Comparative Study of Band-Power Extraction Techniques for Motor Imagery Classification, IEEE Symposium on Computational Intelligence, Cognitive Algorithms, Mind, and Brain, 2011.

[220] G. Dornhege, B. Blankertz, G. Curio, K.R. Muller, Combining features for BCI, Adv. Neural Inf. Process. Syst. (2003) 1139–1146.

[221] F. Lotte, M. Congedo, A. Lecuyer, F. Lamarche, B. Arnaldi, A review of classification algorithms for EEG-based brain-computer interfaces, J. Neural Eng. 4 (2007) R1–R13.

[222] K.R. Muller, C.W. Anderson, G.E. Birch, Linear and nonlinear methods for brain-computer interfaces, IEEE Trans Neural Syst. Rehabil. Eng. 11 (2003) 165–169.

[223] S.G. Mason, A. Bashashati, M. Fatourechi, K.F. Navarro, G.E. Birch, A comprehensive survey of brain interface technology designs, Ann. Biomed. Eng. 35 (2007) 137–169.

[224] B. Blankertz, K.R. Muller, G. Curio, T.M. Vaughan, G. Schalk, J.R. Wolpaw, et al., The BCI competition 2003: progress and perspectives in detection and discrimination of EEG single trials, IEEE Trans. Biomed. Eng. 51 (2004) 1044–1051.

[225] B. Blankertz, K.R. Muller, D.J. Krusienski, G. Schalk, J.R. Wolpaw, A. Schlogl, et al., The BCI competition. III: Validating alternative approaches to actual BCI problems, IEEE Trans. Neural Syst. Rehabil. Eng. 14 (2006) 153−159.

[226] T. Geng, J.Q. Gan, M. Dyson, C.S.L. Tsui, F. Sepulveda, A novel design of 4-class bci using two binary classifiers and parallel mental tasks, Comput. Intell. Neurosci. 2008 (2008) 1−5.

[227] R.A. Fisher, The use of multiple measurements in taxonomic problems, Ann. Eugen. 7 (1936) 179−188.

[228] D.A. Clausi, An analysis of co-occurrence texture statistics as a function of grey level quantization, Can. J. Remote Sens. 28 (2002) 45−62.

[229] R. Scherer, G. Muller, C. Neuper, B. Graimann, G. Pfurtscheller, An asynchronously controlled EEG-based virtual keyboard: improvement of the spelling rate, IEEE Trans. Biomed. Eng. 51 (2004) 979−984.

[230] N.R. Draper, H. Smith, Applied Regression Analysis, Wiley Series in Probability and Statistics, 1998.

[231] R. Fazel-Rezai and W. Ahmad, P300-based Brain-Computer Interface Paradigm Design, Recent Advances in Brain-computer Interface Systems, InTech, 2011.

[232] D.J. Krusienski, E.W. Sellers, F. Cabestaing, S. Bayoudh, D.J. McFarland, T.M. Vaughan, et al., A comparison of classification techniques for the P300 Speller, J. Neural Eng. 3 (2006) 299.

[233] B. Lou, B. Hong, X. Gao, S. Gao, Bipolar electrode selection for a motor imagery based brain-computer interface, J. Neural Eng. 5 (2008) 342−349.

[234] S. Fazli, F. Popescu, M. Danóczy, B. Blankertz, K.R. Müller, C. Grozea, Subject-independent mental state classification in single trials, Neural Networks (2009) 1305−1312.

[235] S.M. Zhou, J.Q. Gan, F. Sepulveda, Classifying mental tasks based on features of higher-order statistics from EEG signals in brain-computer interface, Inf Sci 178 (2008) 1629−1640.

[236] B. Blankertz, M. Tangermann, C. Vidaurre, S. Fazli, C. Sannelli, S. Haufe, et al., The berlin brain-computer interface: non-medical uses of BCI technology, Front. Neurosci. 4 (2010) 1−17.

[237] J. Blumberg, J. Rickert, S. Waldert, A. Schulze-Bonhage, A. Aertsen, C. Mehring, Adaptive classification for brain computer interfaces, 29th Annu. Int. Conf. IEEE Eng. Med. Biol. Soc. (2007) 2536−2539.

[238] C. Vidaurre, A. Schloegl, B. Blankertz, M. Kawanabe, K.R. Müller, Unsupervised adaptation of the LDA classifier for brain-computer interfaces, Proc. 4th Int. Brain-Comput. Inter. Workshop Training Course (2008) 122−127.

[239] B.E. Boser, I.M. Guyon, V.N. Vapnik, A Training Algorithm for Optimal Margin Classifiers Proceedings of the 5th Annual ACM Workshop on Computational Learning Theory, ACM Pres, Pittsburgh, Pittsburgh, PA, 1992.

[240] K. Deng. OMEGA: On-line memory-based general purpose system classifier. Ph.D. Thesis, Carnegie Mellon University, 1998.

[241] W.D. Stirling, Iteratively reweighted least squares for models with a linear part, Appl. Stat. (1984) 7−17.

[242] A. Subasi, E. Erçelebi, Classification of EEG signals using neural network and logistic regression, Comput. Methods Programs Biomed. 78 (2005) 87−99.

[243] J. Q. Gan, Self-adapting BCI based on unsupervised learning, in 3rd International Workshop on Brain-Computer Interfaces, pp. 50−51, 2006.

[244] A. Llera, M. van Gerven, V. Gómez, O. Jensen, H. Kappen, On the use of interaction error potentials for adaptive brain computer interfaces, Neural Networks (2011) 1120−1127.

[245] G. Schalk, J.R. Wolpaw, D.J. McFarland, G. Pfurtscheller, EEG-based communication: presence of an error potential, Clin. Neurophysiol. 111 (2000) 2138−2144.

[246] P.M. Fitts, The information capacity of the human motor system in controlling the amplitude of movement, J. Exp. Psychol. Gen. 121 (1992) 262−269.

[247] A. Satti, Fast Adaptive Signal Processing for Intelligent Multistate Self-Paced BCIs, Ph. D. Thesis, University of Ulster, 2011.

[248] D.J. Krusienski, M. Grosse-Wentrup, F. Galán, D. Coyle, K.J. Miller, E. Forney, et al., Critical issues in state-of-the-art brain-computer interface signal processing, J. Neural Eng. 8 (2011) 025002.

[249] B. Blankertz, G. Dornhege, C. Schafer, R. Krepki, J. Kohlmorgen, K.R. Muller, et al., Boosting bit rates and error detection for the classification of fast-paced motor commands based on single-trial EEG analysis, IEEE Trans. Neural Syst. Rehabil. Eng. 11 (2003) 127−131.

[250] L.C. Parra, C.D. Spence, A.D. Gerson, P. Sajda, Response error correction - a demonstration of improved human-machine performance using real-time EEG monitoring, IEEE Trans. Neural Syst. Rehabil. Eng. 11 (2003) 173−177.

[251] M. Falkenstein, J. Hoormann, S. Christ, J. Hohnsbein, ERP components on reaction errors and their functional significance: a tutorial, Biol. Psychol. 51 (2000) 87−107.

[252] J. Kennedy, The particle swarm: Social adaptation of knowledge, IEEE Int. Conf. Evol. Comput. (2002) 303−308.

[253] J. Kennedy, R.C. Eberhart, Particle swarm optimization, Proc. IEEE Int. Conf. Neural Networks (1995) 1942−1948.

[254] Y. Shi and R. Eberhart, A modified particle swarm optimizer, in Evolutionary Computation Proceedings, 1998. IEEE World Congress on Computational Intelligence., the 1998 IEEE International Conference on, pp. 69−73, 1998. MIT Press Cambridge, MA, USA.

[255] J. H. Holland, Adaptation in Natural and Artificial Systems, MIT Press, Cambridge, MA, USA, 1975.

[256] K. Tanaka, T. Kurita, F. Meyer, L. Berthouze, T. Kawabe, Stepwise feature selection by cross validation for EEG-based brain computer interface, Int. Joint Conf. Neural Networks (2006) 4672−4677.

[257] F. Faradji, R.K. Ward, G.E. Birch, Toward development of a two-state brain-computer interface based on mental tasks, J. Neural Eng. 8 (2011) 046014.

[258] Y. Shi, R.C. Eberhart, Empirical study of particle swarm optimization, Proc. 1999 Congress on Evol. Comput. (1999).

[259] M. Clerc, J. Kennedy, The particle swarm-explosion, stability, and convergence in a multidimensional complex space, IEEE Trans. Evol. Comput. 6 (2002) 58−73.

[260] B. Obermaier, C. Neuper, C. Guger, G. Pfurtscheller, Information transfer rate in a five-classes brain-computer interface, IEEE Trans. Neural Syst. Rehabil. Eng. 9 (2001) 283−288.

[261] M. Naeem, C. Brunner, R. Leeb, B. Graimann, G. Pfurtscheller, Separability of four-class motor imagery data using independent components analysis, J. Neural Eng. 3 (2006) 208.

[262] E.A. Curran, M.J. Stokes, Learning to control brain activity: a review of the production and control of EEG components for driving brain-computer interface (BCI) systems, Brain Cogn. 51 (2003) 326−336.

[263] A. Kübler, B. Kotchoubey, T. Hinterberger, N. Ghanayim, J. Perelmouter, M. Schauer, et al., The thought translation device: a neurophysiological approach to communication in total motor paralysis, Exp. Brain Res. 124 (1999) 223−232.

[264] M. Palankar, K.J. De Laurentis, R. Alqasemi, E. Veras, R. Dubey, Y. Arbel, et al., Control of a 9-DoF wheelchair-mounted robotic arm system using a P300 brain computer interface: initial experiments, IEEE Int. Conf. Rob. Biomim. (2009) 348−353.

[265] B. Blankertz, G. Dornhege, M. Krauledat, M. Schröder, J. Williamson, R. Murray-Smith, K. R. Müller, The Berlin brain-computer interface presents the novel mental

typewriter hex-o-spell, in Proceedings of the 3rd International Brain-Computer Interface Workshop and Training Course, pp. 108–109, 2006.

[266] B. Blankertz, M. Krauledat, G. Dornhege, J. Williamson, R. Murray-Smith, Advanced human-computer interaction with the Berlin brain-computer interface, in BRAINPLAY 07 Brain-Computer Interfaces and Games Workshop at ACE (Advances in Computer Entertainment), 2007.

[267] C.S. Nam, Y. Li, S. Johnson, Evaluation of P300-based brain-computer interface in real-world contexts, Int. J. Hum. Comput. Interact. 26 (2010) 621–637.

[268] E.M. Mugler, C.A. Ruf, S. Halder, M. Bensch, A. Kubler, Design and implementation of a P300-based brain-computer interface for controlling an internet browser, IEEE Trans. Neural Syst. Rehabil. Eng. 18 (2010) 599–609.

[269] B. Rebsamen, E. Burdet, C. Guan, C.L. Teo, Q. Zeng, M. Ang, et al., Controlling a wheelchair using a BCI with low information transfer rate, 10th IEEE Int. Conf. Rehabil. Rob. (2007) 1003–1008.

[270] C.J. Bell, P. Shenoy, R. Chalodhorn, R.P. Rao, Control of a humanoid robot by a non-invasive brain-computer interface in humans, J. Neural Eng. 5 (2008) 214–220.

[271] M. Krauledat, G. Dornhege, B. Blankertz, G. Curio, K.R. Müller, The Berlin brain-computer interface for rapid response, Biomed. Tech. 49 (2004) 61–62.

[272] B. Blankertz, G. Dornhege, M. Krauledat, K.R. Müller, V. Kunzmann, F. Losch, et al., The Berlin Brain-Computer Interface: EEG-based communication without subject training, Hand 3 (2006) C4.

[273] B. Blankertz, M. Krauledat, G. Dornhege, J. Williamson, R. Murray-Smith, K. Muller, A note on brain actuated spelling with the Berlin brain-computer interface, Lecture Notes Comp. Sci. 4555 (2007) 759.

[274] B. Blankertz, G. Dornhege, M. Krauledat, K.R. Müller, G. Curio, The non-invasive Berlin brain-computer interface: Fast acquisition of effective performance in untrained subjects, Neuroimage 37 (2007) 539–550.

[275] K.R. Müller, B. Blankertz, Toward noninvasive brain-computer interfaces, IEEE Signal Process. Mag. 23 (2006) 125–128.

[276] K.R. Müller, M. Tangermann, G. Dornhege, M. Krauledat, G. Curio, B. Blankertz, Machine learning for real-time single-trial EEG-analysis: From brain-computer interfacing to mental state monitoring, J. Neurosci. Methods 167 (2007) 82–90.

[277] B. Rebsamen, E. Burdet, C. Guan, H. Zhang, C.L. Teo, Q. Zeng, et al., A brain-controlled wheelchair based on P300 and path guidance, Proc. IEEE/RAS-EMBS Int. Conf. Biomed. Rob. Biomechatronics (2006).

[278] B. Rebsamen, C.L. Teo, Q. Zeng, M.H. Ang Jr, E. Burdet, C. Guan, et al., Controlling a wheelchair indoors using thought, IEEE Intell. Syst. 22 (2007) 18–24.

[279] D. Vanhooydonck, E. Demeester, M. Nuttin, H. Van Brussel, Shared control for intelligent wheelchairs: An implicit estimation of the user intention, Proc. 1st Int. Workshop Adv. Serv. Rob. (ASER'03) (2003) 176–182.

[280] F.O. Flemisch, C.A. Adams, S.R. Conway, K.H. Goodrich, M.T. Palmer, P.C. Schutte, The H-Metaphor as a guideline for vehicle automation and interaction, NASA Tech. Rep. Server (2003).

[281] L. Tonin, R. Leeb, M. Tavella, S. Perdikis, J.R. Millán, The role of shared-control in BCI-based telepresence, Proc. 29th A. Int. Conf. IEEE Syst. Man Cybern. Soc (2010).

[282] P. Johan, J.R. del Millán, V. Gerolf, L. Eileen, G. Ferran, F. Pierre, et al., Adaptive shared control of a brain actuated simulated wheelchair, Proc. 2007 IEEE 10th Int. Conf. Rehabil. Rob. (2007) 408–414.

[283] M. Nuttin, D. Vanhooydonck, E. Demeester, V. Brussel, Selection of suitable human-robot interaction techniques for intelligent wheelchairs, 11th IEEE Int. Workshop Rob. Human Inter. Commun. (2002) 146–151.

[284] B. Obermaier, G. Muller, G. Pfurtscheller, Virtual keyboard controlled by spontaneous EEG activity, IEEE Trans. Neural Syst. Rehabil. Eng. 11 (2003) 422–426.

[285] J.R. Wolpaw, D.J. McFarland, G.W. Neat, C.A. Forneris, An EEG-based brain-computer interface for cursor control, Electroencephalogr. Clin. Neurophysiol. 78 (1991) 252–259.

[286] J.R. Wolpaw, H. Ramoser, D.J. McFarland, G. Pfurtscheller, EEG-based communication: improved accuracy by response verification, IEEE Trans. Rehabil. Eng. 6 (1998) 326–333.

[287] P. Meinicke, M. Kaper, F. Hoppe, M. Heumann, H. Ritter, Improving transfer rates in brain computer interfacing: a case study, Adv. Neural Inf. Process. Syst. (2003) 1131–1138.

[288] M. Fatourechi, S.G. Mason, G.E. Birch, R.K. Ward, Is information transfer rate a suitable performance measure for self-paced brain interface systems?, IEEE Int. Symp. Signal Process. Inf. Technol. (2006) 212–216.

[289] G. Townsend, B. Graimann, G. Pfurtscheller, A comparison of common spatial patterns with complex band power features in a four-class BCI experiment, IEEE Trans. Biomed. Eng. 53 (2006) 642–651.

[290] G.M. Foody, On the compensation for chance agreement in image classification accuracy assessment, Photogramm. Eng. Remote Sens. 58 (1992) 1459–1460.

[291] J. Kronegg, S. Voloshynovskiy, T. Pun, Analysis of bit-rate definitions for brain-computer interfaces, Int. Conf. Human-Comput. Interact. (HCI'05) (2005). Las Vegas, Nevada, USA.

[292] I.S. MacKenzie, Fitts' law as a research and design tool in human-computer interaction, Hum. Comput. Interact. 7 (1992) 91–139.

[293] C.E. Shannon, A mathematical theory of communication, ACM SIGMOBILE Mobile Comput. Commun. Rev. 5 (2001) 3–55.

[294] J. Decety, M. Jeannerod, Mentally simulated movements in virtual reality: does Fitt's law hold in motor imagery? Behav. Brain Res 72 (1995) 127–134.

[295] E.A. Felton, R.G. Radwin, J.A. Wilson, J.C. Williams, Evaluation of a modified Fitts law BCI target acquisition task in able and motor disabled individuals, J. Neural Eng. 6 (2009) 056002.

[296] C. Choi, C. Kim, J. Kim, Comparison study of biosignal based computer interfaces based on Fitts' law paradigm, Proc. 17th IFAC World Congress 17 (2008).

[297] B. Dal Seno, M. Matteucci, L. Mainardi, The utility metric: A novel method to assess the overall performance of discrete brain-computer interfaces, IEEE Trans. Neural Syst. Rehabil. Eng. 18 (2010) 28.

[298] I. Iturrate, J.M. Antelis, A. Kubler, J. Minguez, A noninvasive brain-actuated wheelchair based on a P300 neurophysiological protocol and automated navigation, IEEE Trans. Rob. 25 (2009) 614–627.

[299] K. Choi, A. Cichocki, Control of a wheelchair by motor imagery in real time, Intell. Data Eng. Autom. Learn. − IDEAL 2008 (2008) 330–337.

[300] B. Rebsamen, C. Guan, H. Zhang, C. Wang, C. Teo, M.H. Ang, et al., A brain controlled wheelchair to navigate in familiar environments, IEEE Trans. Neural Syst. Rehabil. Eng. 18 (2010) 590–598.

[301] N. Zettili, Quantum Mechanics: Concepts and Applications, John Wiley & Sons Inc, 2009.

[302] R. Shankar, Principles of Quantum Mechanics, Springer, 1994.

[303] D.A.B. Miller, Quantum Mechanics for Scientists and Engineers, Cambridge Univ Pr, 2008.

[304] T.R. Taha, M.I. Ablowitz, Analytical and numerical aspects of certain nonlinear evolution equations. II. Numerical, nonlinear Schrödinger equation, J. Comput. Phys. 55 (1984) 203–230.

[305] DARPA Neural Network Study (US), AFCEA International Press, 1988.

[306] S. Sivanandam, Introduction to Neural Networks using MATLAB 6.0, Tata McGraw-Hill Education, 2006.

[307] P.A.M. Dirac, Bakerian lecture. The physical interpretation of quantum mechanics, Proc. R. Soc. Lond. A Math. Phys. Sci. 180 (1942) 1.

[308] G.G. Globus, K.H. Pribram, G. Vitiello, Brain and Being: At the Boundary Between Science, Philosophy, Language and Arts, John Benjamins Publishing Company, 2004.

[309] E. Pessa, Brain-computer interfaces and quantum robots, Arxiv Preprint arXiv 0909 (2009) 1508.

[310] E. Conte, A.Y. Khrennikov, O. Todarello, A. Federici, L. Mendolicchio, J.P. Zbilut, Mental states follow quantum mechanics during perception and cognition of ambiguous figures, Arxiv Preprint arXiv 0906 (2009) 4952.

[311] J.A. de Barros, P. Suppes, Quantum mechanics, interference, and the brain, J. Math. Psychol. 53 (2009) 306–313.

[312] J.R. Busemeyer, Z. Wang, J.T. Townsend, Quantum dynamics of human decision-making, J. Math. Psychol. 50 (2006) 220–241.

[313] R.S. Bucy, Linear and nonlinear filtering, Proc. IEEE 58 (1970) 854–864.

[314] K.H. Přibram, Rethinking Neural Networks: Quantum Fields and Biological Data, Lawrence Erlbaum, 1993.

[315] R.L. Dawes, Quantum neurodynamics: neural stochastic filtering with the Schroedinger equation, Int. Joint Conf. Neural Netw. (1992).

[316] R.P. Feynman, Quantum mechanical computers, Found. Phys. 16 (1986) 507–531.

[317] J.R. Searle, Minds, Brains, and Science, Harvard Univ Pr, 1984.

[318] L. Behera, B. Sundaram, Stochastic filtering and speech enhancement using a recurrent quantum neural network, Proc. Int. Conf. Int. Sens. Inf. Process. (2004) 165–170, ICISIP-2004, pp. 165.

[319] L. Behera, I. Kar, A.C. Elitzur, A recurrent quantum neural network model to describe eye tracking of moving targets, Found. Phys. Lett. 18 (2005) 357–370.

[320] L. Behera, I. Kar, Quantum stochastic filtering, IEEE International Conference on Systems, Man and Cybernetics, 2005.

[321] J. Aldrich, RA Fisher and the making of maximum likelihood 1912–1922, Stat. Sci. 12 (1997) 162–176.

[322] S. Haykin, Neural Networks, A Comprehensive Foundation. Prentice Hall PTR, Upper Saddle River, NJ, USA, 1994.

[323] G.C. Onwubolu, B.V. Babu, New Optimization Techniques in Engineering, Springer Verlag, 2004.

[324] H. Muehlenbein, T. Mahnig, Foundations of Real-World Intelligence, chapter –Evolutionary Computation and Beyond, CSLI Publications, 2001.

[325] L. Behera, I. Kar, A. C. Elitzur, Recurrent Quantum neural network and its applications, The Emerging Physics of Consciousness, Springer, 2006, pp. 327–350.

[326] L. Behera, S. Bharat, S. Gaurav, A. Manish, A recurrent network model with neurons activated by Schroedinger wave equation and its application to stochastic filtering, 9th International Conference on High-Performance Computing, Workshop on Soft Computing, 2002.

[327] V. Gandhi, G. Prasad, D. Coyle, L. Behera, T.M. McGinnity, Quantum neural network based EEG filtering for a brain-computer interface, IEEE Trans. Neural Networks Learn. Syst. 25 (2014) 278–288.

[328] J. Crank, P. Nicolson, A practical method for numerical evaluation of solutions of partial differential equations of the heat-conduction type, Math. Proc. Camb. Philos. Soc. (1947) 50–67.

[329] J. Scheffel, Does Nature Solve Differential Equations? TRITA-ALF-2002–02, Royal Institute of Technology, Stockholm, Sweden, 2002.

[330] Evolving of the wave packet [Online]. Available: <http://isrc.ulster.ac.uk/images/ stories/Staff/BCI/Members/VGandhi/Video_PhysicalRobotControl/wavepacket_ evolves_according_to_swe.mp4> (accessed 16.07.14).

[331] R. Penrose, Shadows of the Mind: A Search for the Missing Science of Consciousness, Oxford University Press, USA, 1996.

[332] G. Dorffner, Neural Networks and a New Artificial Intelligence, Itp-Media, 1997.

[333] R. Ron-Angevin, F. Velasco-Alvarez, S. Sancha-Ros, L. da Silva-Sauer, A two-class self-paced BCI to control a robot in four directions, IEEE Int. Conf. Rehabil. Rob. (2011) 1–6.

[334] A.R. Satti, D. Coyle, G. Prasad, Self-paced brain-controlled wheelchair methodology with shared and automated assistive control, pp. 1–8 IEEE Symposium on Computational Intelligence, Cognitive Algorithms, Mind, and Brain (CCMB), 2011.

[335] Player/Stage/Gazebo: Free software tools for robot and sensor applications [Online]. Available: <http://playerstage.sourceforge.net/> (accessed 16.07.14).

[336] Schunk Robots [Online]. Available: <http://www.schunk.com/schunk_files/ attachments/ModularRobotics_2010–06_EN.pdf> (accessed 16.07.14).

[337] J. Postel, RFC 768: user datagram protocol, Network Inf. Center (1980).

[338] V. Gandhi, G. Prasad, D. Coyle, L. Behera, T.M. McGinnity, EEG based mobile robot control through an adaptive brain-robot interface, IEEE Trans. Syst. Man Cybern.: Syst. (2014) (IEEE early access)

[339] C. Brunner, R. Leeb, G. R. Müller Putz, A. Schlögl and G. Pfurtscheller, BCI Competition 2008 – Graz data set A, 2009.

[340] R. Leeb, C. Brunner, G. R. Müller-Putz, A. Schlögl and G. Pfurtscheller, BCI Competition 2008 – Graz data set B, 2009.

[341] B. Blankertz, BCI competitions IV. Available online at: <http://www.bbci.de/ competition/iv/>, 2008 (accessed 16.07.14).

[342] O. Väisänen, Multichannel EEG Methods to Improve the Spatial Resolution of Cortical Potential Distribution and the Signal Quality of Deep Brain Sources, Tampereen Teknillinen Yliopisto, vol. 741, Julkaisu-Tampere University of Technology. Publication, 2009.

[343] S. Shahid, G. Prasad, Bispectrum-based feature extraction technique for devising a practical brain-computer interface, J. Neural Eng. 8 (2011) 025014.

[344] Robot Control Through Motor Imagery [Online]. Available: <http://isrc.ulster.ac. uk/Staff/VGandhi/VideoRobotControlThroughMI> (accessed 16.07.14).

[345] C. Guger, G. Krausz and G. Edlinger, Brain-computer interface control with dry EEG electrodes, g.Tec Available online on: <http://www.gtec.at/content/down-load/5152/41756/version/1/> (accessed 16.07.14).

[346] B. Gerkey, R. Vaughan, A. Howard and N. Koenig, The Player/Stage Project, Hosted at <http://playerstage.Sourceforge.Net>, 2003 (accessed 16.07.14).

[347] T. Geng, M. Dyson, C. Tsui, J.Q. Gan, A 3-class asynchronous BCI controlling a simulated mobile robot, 29th Annu. Int. Conf. IEEE Eng. Med. Biol. Soc. (2007) 2524–2527.

[348] R. Leeb, H. Sagha, R. Chavarriaga, J.R. Millán, Multimodal fusion of muscle and brain signals for a hybrid-BCI, Proc. Int. Conf. IEEE Eng. Med. Biol. Soc. (2010).

[349] B.Z. Allison, C. Brunner, V. Kaiser, G.R. Müller-Putz, C. Neuper, G. Pfurtscheller, Toward a hybrid brain-computer interface based on imagined movement and visual attention, J. Neural Eng. 7 (2010) 026007.

[350] R. Scherer, G.R. Müller-Putz, G. Pfurtscheller, Self-initiation of EEG-based brain-computer communication using the heart rate response, J. Neural Eng. 4 (2007) L23.

[351] S. Shahid, G. Prasad, R.K. Sinha, On fusion of heart and brain signals for hybrid BCI, 5th Int. IEEE/EMBS Conf. Neural Eng. (NER) (2011).

[352] G. Pfurtscheller, B.Z. Allison, C. Brunner, G. Bauernfeind, T. Solis-Escalante, R. Scherer, et al., The hybrid BCI, Front. Neurosci. 4 (2010) 1–11.

[353] G. Pfurtscheller, T. Solis-Escalante, R. Ortner, P. Linortner, G.R. Muller-Putz, Self-paced operation of an SSVEP-based orthosis with and without an imagery-based brain switch: a feasibility study towards a hybrid BCI, IEEE Trans. Neural Syst. Rehabil. Eng. 18 (2010) 409–414.

[354] L. Simmons, N. Sharma, J.C. Baron, V.M. Pomeroy, Motor imagery to enhance recovery after subcortical stroke: who might benefit, daily dose, and potential effects, Neurorehabil. Neural Rep. 22 (2008) 458–467.

[355] C. Guger, S. Daban, E. Sellers, C. Holzner, G. Krausz, R. Carabalona, et al., How many people are able to control a P300-based brain-computer interface (BCI)?, Neurosci Lett 462 (2009) 94–98.

[356] J.R. Millán, R. Rupp, G.R. Müller-Putz, R. Murray-Smith, C. Giugliemma, M. Tangermann, et al., Combining brain-computer interfaces and assistive technologies: state-of-the-art and challenges, Front. Neurosci. 4 (2010) 1–15.

[357] P.W. Ferrez, J.R. Millán, Simultaneous real-time detection of motor imagery and error-related potentials for improved BCI accuracy, Proc. 4th Int. Brain-Computer Int. Workshop & Training Course (2008) 197–202.

[358] C. Brunner, B.Z. Allison, D.J. Krusienski, V. Kaiser, G.R. Müller-Putz, G. Pfurtscheller, et al., Improved signal processing approaches in an offline simulation of a hybrid brain-computer interface, J. Neurosci. Methods 188 (2010) 165–173.

[359] G. Pfurtscheller, T. Solis-Escalante, Could the beta rebound in the EEG be suitable to realize a brain switch? Clin. Neurophysiol. 120 (2009) 24–29.

[360] M. Danoczy, S. Fazli, C. Grozea, K.R. Müller, F. Popescu, Brain2robot: A grasping robot arm controlled by gaze and asynchronous EEG BCI, Proc. 4th Int. BCI Workshop & Train. Course (2008).

[361] T.O. Zander, M. Gaertner, C. Kothe, R. Vilimek, Combining eye gaze input with a brain-computer interface for touchless human-computer interaction, Intl, J. Human-Computer Interact. 27 (2010) 38–51.

[362] R.D. Wright, L.M. Ward, Orienting of Attention, Oxford University Press, USA, 2008.

[363] A.R. Hunt, A. Kingstone, Covert and overt voluntary attention: linked or independent? Cogn. Brain Res. 18 (2003) 102–105.

[364] S. Mason, G. Birch, A brain-controlled switch for asynchronous control applications, IEEE Trans. Biomed. Eng. 47 (2000) 1297–1307.

[365] L. Mutch, E. Alberman, B. Hagberg, K. Kodama, M.V. Perat, Cerebral palsy epidemiology: Where are we now and where are we going? Develop. Med. Child Neurol. 34 (1992) 547–551.

[366] J.R. Daube, Electrodiagnostic studies in amyotrophic lateral sclerosis and other motor neuron disorders, Muscle Nerve 23 (2000) 1488–1502.

[367] S.J. King, M.M. Duke, B.A. O'Connor, Living with amyotrophic lateral sclerosis/motor neurone disease (ALS/MND): decision-making about 'ongoing change and adaptation, J. Clin. Nurs. 18 (2009) 745–754.

[368] T.F. Collura, History and evolution of electroencephalographic instruments and techniques, J. Clin. Neurophysiol. 10 (1993) 476–504.

[369] C.M. Stinear, W.D. Byblow, M. Steyvers, O. Levin, S.P. Swinnen, Kinesthetic, but not visual, motor imagery modulates corticomotor excitability, Exp. Brain Res. 168 (2006) 157–164.

[370] K.J. Miller, G. Schalk, E.E. Fetz, M. Den Nijs, J.G. Ojemann, R.P.N. Rao, Cortical activity during motor execution, motor imagery, and imagery-based online feedback, Proc. Nat. Acad. Sci. 107 (2010) 4430–4435.

[371] C. Kranczioch, S. Mathews, P.J.A. Dean, A. Sterr, On the equivalence of executed and imagined movements: Evidence from lateralized motor and non-motor potentials, Hum. Brain Mapp. 30 (2009) 3275–3286.

[372] M. Lotze, U. Halsband, Motor imagery, J. Physiol. Paris 99 (2006) 386–395.

[373] A.J. Szameitat, A. McNamara, S. Shen, A. Sterr, Neural activation and functional connectivity during motor imagery of bimanual everyday actions, PloS One 7 (2012) e38506.

[374] N. Picard, P.L. Strick, Activation of the supplementary motor area (SMA) during performance of visually guided movements, Cereb. Cortex 13 (2003) 977–986.

[375] C.E. Shannon, W. Weaver, The Mathematical Theory of Communication, University of Illinois Press Urbana, 1962.

[376] J.R. Pierce, An Introduction to Information Theory: Symbols, Signals & Noise, Dover Pubns, 1980.

[377] J. Cohen, A coefficient of agreement for nominal scales, Educ. Psychol. Meas. 20 (1960) 37–46.

[378] H. Mühlenbein, The equation for response to selection and its use for prediction, Evol. Comput. 5 (1997) 303–346.

[379] H. Mühlenbein, G. Paass, From recombination of genes to the estimation of distributions I. Binary parameters, Parallel Problem Solving Nat. – PPSN IV (1996) 178–187.

[380] J. Brownlee, Clever algorithms: nature-inspired programming recipes, Lulu. Com (2011).

[381] Pioneer 3D-X [Online]. Available: <http://www.mobilerobots.com/ResearchRobots/PioneerP3DX.aspx> (accessed 16.07.14).

[382] T.O. Zander, M. Lehne, K. Ihme, S. Jatzev, J. Correia, C. Kothe, et al., A dry EEG-system for scientific research and brain-computer interfaces, Front. Neurosci. 5 (2011) 1–10.

[383] G. Gargiulo, P. Bifulco, R.A. Calvo, M. Cesarelli, C. Jin, A. McEwan, et al., Non-invasive electronic biosensor circuits and systems, edited by V.S. Somerset, Intell. Biosens. Tech (2010) 123–146.

Index

Note: Page numbers followed by "*f*", "*t*" and "*b*" refer to figures, tables and boxes, respectively.

Printed in the United States
By Bookmasters